LEADERSHIP AND LOCAL POWER IN EUROPEAN RURAL DEVELOPMENT

T0253394

Leadership and Local Power in European Rural Development

Edited by

KEITH HALFACREE
University of Wales, Swansea

IMRE KOVÁCH
Institute for Political Sciences, Budapest

RACHEL WOODWARD
University of Newcastle

Routledge
Taylor & Francis Group

LONDON AND NEW YORK

First published 2002 by Ashgate Publishing

Published 2017 by Routledge
2 Park Square, Milton Park, Abingdon, Oxfordshire OX14 4RN
711 Third Avenue, New York, NY 10017, USA

First issued in paperback 2017

Routledge is an imprint of the Taylor & Francis Group, an informa business

British Library Cataloguing in Publication Data
Halfacree, Keith
  Leadership and local power in European rural development. -
  (Perspectives on rural policy and planning)
  1.Rural development - Europe 2. Political leadership -
  Europe 3. Local government - Europe
  I.Title II.Kovach, Imre III.Woodward, Rachel
  307.1'212'094

Library of Congress Control Number: 2001095872

ISBN 13: 978-1-138-26384-0 (pbk)
ISBN 13: 978-0-7546-1581-1 (hbk)

Contents

List of Figures

List of Tables

List of Contributors

Titus Bahner is a lecturer in the Faculty of Economics at the University of Witten-Herdecke in Germany.

Nina Gunnerud Berg is an Associate Professor in the Department of Geography at the Norwegian University of Science and Technology (NTNU) in Trondheim, Norway.

Chris Curtin is a professor and head of the Department of Political Science and Sociology at the National University of Ireland, Galway.

Thomas Dax is the Deputy Director of the Federal Institute for Less-Favoured and Mountainous Areas in Vienna, Austria.

Philippe Gajewski is *Doctorant* at the Laboratoire des Dynamiques sociales et recomposition des espaces, University of Paris X, France.

Fernando Garrido is a Researcher in the Andalusia Institute for Social Studies (IESA-CSIC) in Córdoba, Spain.

Henri Goverde is Associate Professor in Public Administration in the Faculty of Management Sciences at the University of Nijmegen, and Professor of Political Science at Wageningen University, both in the Netherlands.

Henk de Haan is a professor in the Rural Sociology Group, Department of Social Sciences at Wageningen University in the Netherlands.

Keith Halfacree is a Lecturer in the Geography Department of the University of Wales, Swansea.

Martin Hebertshuber is a freelance researcher and member of Helix Research and Consulting, based in Salzburg, Austria.

Torsti Hyyryläinen is a Senior Researcher at the Mikkeli Institute for Rural Research and Training at the University of Helsinki in Finland.

Imre Kovách is Head of Department at the Institute for Political Sciences at the Hungarian Academy of Sciences in Budapest, Hungary.

Lutz Laschewski is a Researcher at the Institute for Agricultural Economics and Social Sciences in the Agricultural and Horticultural Faculty of the Humboldt University in Berlin, Germany.

Hans Kjetil Lysgård is a researcher at the Agder Research Foundation in Kristiansand, Norway.

Nicole Mathieu is Director of Research at the CNRS, working in the Laboratoire des Dynamiques sociales et recomposition des espaces, University of Paris X, France.

José Ramón Mauleón is a Lecturer in the Department of Sociology 2, University of the Basque Country in Bilbao, Spain.

Eduardo Moyano is the Deputy Director of the Andalusia Institute for Social Studies (IESA-CSIC) in Córdoba, Spain.

Parto Teherani-Krönner is head of the Gender Research Unit at the Institute for Agricultural Economics and Social Sciences in the Agricultural and Horticultural Faculty of the Humboldt University in Berlin, Germany.

Eero Uusitalo is a Councillor for Rural Development in the Finnish Ministry of Agriculture and Forestry.

Tony Varley is a lecturer in the Department of Political Science and Sociology at the National University of Ireland, Galway.

Rachel Woodward is a Senior Research Associate in the Centre for Rural Economy at the University of Newcastle upon Tyne in the UK.

Acknowledgements

This volume is the outcome of a series of meetings between the contributors, meeting as Working Group 5: Leadership and Local Power. These meetings were financed under the European Commission's COST Action A12 on Rural Innovation. The contributors would like to acknowledge the support granted to the group under this programme, which enabled these meetings to take place.

Rachel Woodward would like to thank Emma Paisley and Caroline Faddy in the Department of Agricultural Economics and Food Marketing, University of Newcastle, for their assistance in the preparation of the manuscript.

The editors would like to thank Pat FitzGerald for her typesetting assistance in the preparation of the final manuscript.

Chapter One

Introduction:
A Comparative European Perspective

Keith Halfacree, Imre Kovách and Rachel Woodward

Rural Restructuring and the Reconfiguration of Power Relations

It is a truism that contemporary processes of economic, social, political and cultural restructuring, many of which are global in origin, are having pronounced impacts on the form and function of rural areas across Europe. Whilst these processes may be broadly similar in origin, their outcomes are highly differentiated across space. This diversity of outcomes reflects local, regional and national responses to the challenge of restructuring, visible in the differential geographical impact of economic and social changes, the policies that have been developed to meet these challenges and the ways in which different national traditions have responded to processes of restructuring. The challenges of restructuring are particularly problematic for Europe's rural areas, given the scale of changes in modes of agricultural production and the specific problems economic and social trends have raised for rural localities. Indeed, the importance of both agricultural and rural development policy initiatives within the European Union (EU) is indicative of the significance and specificity of the problems of rural restructuring. A point of departure for this volume is the recognition both of the significance of pan-European economic and social changes, and of the significance of such changes for rural areas. Its starting-point is the tension between diversity and homogeneity, between local responses and global processes.

These broad processes of economic and social change in Europe's rural areas have consequences for local configurations of power, and for local political responses to these broad processes. The study of these configurations and responses is important. These economic and social changes have produced pronounced inequalities across space which in turn have prompted responses (and perceptions) at national and EU levels of the need for economic and social intervention or development. (Marsden, 1998 and 1999; Bell and Lowe, 1998; Starosta et al., 1999). One policy challenge across the EU is the production of a coherent 'Europe of the Regions' which strikes a balance between the specificity of local conditions and experiences, and normative desires for economic prosperity and stability across Europe (Woods, 1998). Moreover, with the retreat from the top-down (exogenous) model of development at national and EU levels (for political, pragmatic and ideological reasons), local responses have gained in significance. As Granberg and Kovách (1998) and Tovey (1998) argue, local actors play a key role in determining responses to restructuring, and the identification of such actors and their roles constitutes a specific research question in European rural studies. Furthermore, certain policy shifts (such as the adoption by the EU of the model of endogenous rural development) have accorded greater (and ever-increasing) significance to the work of local actors and the operation of local networks of power (Murdoch and Marsden, 1994; Shucksmith, 2000). Quite simply, the local is of increasing significance in EU development programmes; LEADER, for example, is viewed as a template for future EU development policy programmes (Ray, 2000). With this institutional emphasis on endogenous development comes an imperative to investigate the influences on local political configurations, power relations and the way they shape local responses to global processes (de Haan and van der Ploeg, 1994; Ray, 1999a, 1999b; Goverde et al., 2000). Who controls and who benefits from changing power relations as a consequence of development initiatives (Bruckmeier, 2000; Buller, 2000; Kovách, 2000; Osti, 2000; Ray, 2000)? Finally, the study of local configurations of rural power and local responses to restructuring are of wider interest because of what they tell us about cultural issues such as the construction of national and regional identities (Woods, 1997). These identities are influenced across Europe by rural images and constructs; it is pertinent, therefore, to see how these identities are developed and mobilised at the local level within a rural context (Halfacree, 1993; Halfacree, 1995; Hoggart et al., 1995; Pratt, 1996; Frouws, 1998). This volume therefore constitutes a contribution to a growing body of research on local responses, local actors, questions of leadership, power, autonomy and control in the evolution of responses to wider economic and social restructuring across rural Europe. Its purpose is to examine the diversity of these local political responses with specific reference to the implications of these responses for local configurations of power in rural areas.

The contributions to this edited volume have been guided by the following observations. First, many of the economic processes of restructuring affect all European countries; these include the move towards post-productivism and associated changes in agriculture, reform of the EU's Common Agricultural

Policy (CAP) and consequences of negotiations on the General Agreement on Tariffs and Trade (GATT). There are also common social trends in operation, including age- and class-selective in- and out-migration between urban and rural areas and processes of urbanisation (Halfacree, 1994; Boyle and Halfacree, 1998; Shucksmith and Chapman, 1998). Different contributors have chosen to focus on different processes of change, reflecting both national situations and the disciplinary backgrounds of individual contributors. Second, following from this, these outcomes of change will be shaped by national differences in the structure of national, regional and local governance, the location of political control of rural development and agriculture within administrative and governmental systems, the availability and use of EU financial assistance under the Structural Funds and Community Initiatives (including LEADER) by governments within member states, and so on. Again, contributors focus on what is important in national contexts and the factors which speak to their individual academic and political concerns. Third, this volume recognises the importance of cultural factors in shaping political responses to rural restructuring. These include differences in the meaning of 'rural' and national cultural constructions of rurality (as well as normative definitions of rural) and how the rural is imagined, produced, represented and conceptualised in different cultural contexts across Europe (Cloke and Little, 1996; Ray, 1998). Contributors to this volume all recognise the ways in which constructions of the rural in different national contexts shape political responses to rural restructuring, and how local power finds expression through the dominance of certain constructions of rurality over others. Underpinning many of the contributions to this volume is the observation that the ways in which rural problems are defined and solutions developed owes as much to cultural conceptions of what the rural is and should be, as to the economic context in which such problems appear or the policy mechanisms developed for their solution.

A Framework for Comparative Study

This volume is structured around individual national contributions, with the choice of countries for comparison dictated by membership of the group behind the production of this volume. The group comprised members of a Working Group on Leadership and Local Power, which was brought together under the EU's COST Action A12 on Rural Innovation (COST being an initiative for collaborative research endeavour). The Working Group had participants from established EU member states (Germany, France, Ireland, the UK, the Netherlands and Spain), more recent EU members (Finland and Austria), one aspirant EU member (Hungary) and one non-EU member (Norway). Each chapter has been written by a leading academic, working in disciplines ranging from rural sociology, to geography and demography, to political science. All comment on their own national contexts.

A challenge in producing this volume was to bring together the diversity of experience and disciplinary outlook. All contributors were given a brief to examine a set of questions with reference to their own national contexts. The first of these was: who sets the agenda in the evolution of rural development policy at the national level, and how indeed is 'rural' defined within different national contexts? The purpose of this question was to tease out how national agendas are set within their national administrative and historical contexts. The second question concerned the extent to which established and emerging social divisions or social groups affect the evolution and administration of rural development policies and programmes in different European countries. How are these divisions manifest or given a material reality? This includes consideration of the roles of new and old elites, the (often conflicting) priorities of 'insiders' and 'outsiders', 'locals' and 'newcomers', issues such as regionalism and questions of gender relations. The third question was: who controls and who benefits from rural development policies at local, regional and national levels? Are the benefits uniform, or confined to specific groups? If uneven, is this politically endorsed or alternatively perceived as unfair? The fourth question concerned any consistency in emerging patterns of leadership and local power within rural Europe, and was aimed at teasing out any common features of elites which have emerging during processes of rural restructuring. In addressing these questions, the chapters have been written to a broadly similar brief given to each individual contributor. Each sets out the social and economic context for rural areas in the countries under question. The administrative context is introduced, to give the unfamiliar reader a grasp of the structures in which administrative decisions relating to rural development and rural politics are made. The political control of rural development is addressed, often with reference to case studies of specific national and regional programmes for rural development. Finally, the evidence for different discourses or cultural constructions of rurality is assessed where appropriate.

The Chapters

The volume opens with an analysis by Chris Curtin and Tony Varley of changing configurations of power and leadership in rural Ireland. Drawing on an analysis of recent policy changes, they examine the rupture to old established patterns of rural leadership in Ireland. This change is conceptualised in terms which distinguish between 'power over' and 'power to'. They argue that the prominent pattern of rural leadership in the Irish past, and the local power structures which this reflected and reproduced, was rooted in 'power over' relationships of domination and subordination. The separation of this pattern from more recently emerging patterns of leadership associated with a partnership approach is a consequence not of the disappearance of the 'power over' dimension, but rather a reconfiguring of the relationship between the dimensions of 'power over' and 'power to'.

This is followed by a discussion by Henri Goverde and Henk de Haan of the politics of rural development in the Netherlands. In the most densely populated country in Europe, the authors argue that discussions of the politics of rural development should be situated within the context of national debates over the definition of rural. These debates entail the possible elimination of rural-urban differences, yet are coupled with what the authors term a 'rural renaissance', a re-orientation of the significance of the rural in Dutch society and the assertion of the cultural value of rural areas. The review of administrative structures and control over rural development policy and planning in the Netherlands focuses on how constructions of the rural in policy discourse can be read from national rural planning documentation. The paper then goes on to explore rural discourses in the Netherlands with reference to two specific instances where images and realities of rural life have been drawn into conflict through the activities of different social groups. These conflicts are understood by the authors as competing conceptualisations of rurality – in one case a social conflict between recent and more established residents of a particular village, and in the other an institutional conflict over the precedence which should be granted to either housing development or environmental protection. The authors conclude that both continuity and fragmentation mark the valorisation of the rural in Dutch social and political life.

In a review of the UK experience, Keith Halfacree and Rachel Woodward examine the socioeconomic, administrative and cultural contexts in which rural development programmes have taken place. Reviewing changes to the socioeconomic structure of the UK, they highlight counter-urbanisation as the most significant socioeconomic force affecting rural localities. A review of the administrative structures influencing rural development in England precedes an examination of the broad features of both European and national rural development programmes, with particular emphasis on the consequences of these programmes for changing local power relations. The chapter concludes with a review of recent scholarship on discourses of rurality, showing the dominance of the notion of the rural idyll which reflects deep-seated cultural ideas about what the countryside is, and who it is for.

The situation and outlook for an aspirant EU member, Hungary, is considered by Imre Kovách. The transition from socialism prompted wide-ranging reforms across the Hungarian economy, with radical reforms of the agricultural system. More recently, and in anticipation of Hungary's accession to the EU, a regional/rural development system has been introduced to promote economic development and to work towards the reduction in regional disparities across Hungary. The chapter describes this new system, paying particular attention to the restratification of Hungarian rural society which has brought both the emergence of new rural elites and the exclusion of poorer social groups. The new actors in the rural development system are considered with particular reference to the changing configurations of power which have accompanied this process. The chapter concludes with a consideration of the relationships between contemporary processes of rural restructuring and the historical and contemporary construction of images of rurality.

Rural restructuring, power distribution and leadership at national, regional and local levels in France are examined by Nicole Mathieu and Philippe Gajewski. The authors identify two characteristics defining (rural) France, these being a national self-representation of France as a peasant society despite the small minority actually engaged in agriculture, and a French tradition of decentralised administration through the communes. These two features above all else have shaped the character of rural restructuring in France. The chapter focuses on the influence of debates on the meaning of rurality within policy discourses on development, and on the influences in social and agricultural spheres of emerging policy networks and new social movements in development policy. The authors go on to explore a theorisation of the relationship between local/rural development policies and empowerment, introducing the idea of a rural 'observatory'.

Lutz Laschewski, Parto Teherani-Krönner and Titus Bahner look at rural restructuring in the context of the different experiences of the former West and East Germany in the years following reunification in 1990. Following this, one set of institutions (on the West German model) was put in place across Germany, but operating to different effects in East and West. The authors argue that the post-unification framework of regional and agricultural policies is currently under pressure because of regional heterogeneity and the power of states within the federal system. There is a well-established system of regional development funding and programmes. European funding has been integrated within this system, and the financial effects have been significant for East Germany. The popular Village Renewal scheme East Germany is also discussed. The chapter goes on to examine the impacts of social changes on rural life, including those on the social composition of the rural population, and the concept of the 'regionalised village' in West Germany. This contrasts with the post-war reconstruction of rural East Germany, the experience of which is examined through an assessment of the effect of agricultural industrialisation under East German socialism, and its consequences for local power and gender relations. The final section discusses how rurality as a concept is subsumed within policy discourses on regionality, how cultural and social forces give the rural its meanings, and the regional differences evident in this. The chapter concludes by discussing local power and participation, the retreat from local democracy and mechanisms for regulating rural conflicts, contrasting the experiences of East and West Germany.

Rural restructuring and the effects of rural development policies in Spain are examined by Fernando Garrido, José Mauleón and Eduardo Moyano. The past two decades of rural restructuring have witnessed political changes including democratisation, with administrative decentralisation and EU accession; cultural changes, including a new emphasis on the local and on the environment; and economic changes, including the emergence of new non-farming actors in rural areas. Under a decentralised system of government, regional governments have power over agriculture and rural development within their territories, whilst central government liaises with the EU. Rural areas have seen industrialisation and the out-migration of younger people in the 1980s but more recently there has been a reversal of these trends coupled with a revaluation of rural areas. Spanish rural sociology has been dominated by debates over the existence of a 'true' rural society, and the most appropriate way in which to define this. The chapter looks at policy responses to agricultural and rural economic changes, including LEADER II, Objectives 1 and 5b, and PRODER, a Spanish initiative. The adoption of these policies is judged by the authors to have invigorated and given voice to rural areas across Spain, encouraging the participation of social actors in the development process. The chapter provides a case study of Andalucia to illustrate this. The authors then go on to discuss the emergence of new local identities, social participation, and a managerial and technical structure for development. A significant achievement is judged to be the establishment of a multidisciplinary governmental structure which offers training and experience in rural development. The chapter then goes on to discuss how new actors are providing redefinitions of the rural, highlighting the importance of the urban population in constructing ideas of rurality. The chapter concludes with some observations on the role of development programmes in building social capital in rural Spain.

Thomas Dax and Martin Hebertshuber consider regional and rural development and its influence on leadership and local power in the Austrian context. Their contribution focuses explicitly on administrative and policy responses to rural restructuring. The chapter introduces the administrative structures for the support of rural regions, focusing on 'bottom-up' movements in regional development, before going on to consider the major changes for regional development following the introduction and implementation of the EU Structural Funds, after Austria's EU accession in 1995. The authors argue that these new policy measures have had a discernible influence on the former balance of leadership and local power, and have triggered new dynamics in rural areas. They argue that whilst it is too early to make any definitive assessments of the actual effects of these new dynamics, they identify both positive impacts and (negative) deficiencies in the rural development measures and institutions in Austria.

Another recent EU member, Finland, is considered by Torsti Hyyrylainen and Eero Uusitalo, who consider recent rural restructuring and rural policy in this sparsely populated country. Finland has had a strong tradition of an independent self-sufficient peasantry, and was strongly agrarian until the 1960s. The authors present the period 1989–99 as something of a watershed in Finnish rural policy, and they describe these changes with reference to two village action programmes, LEADER, and POMO (a Finnish supplement to LEADER to encourage local initiatives and sub-regional development). There have been difficulties in the implementation of the programmes because of tensions between central and regional administrations, and the process of local decision-making. The relationship between national rural development policy initiatives and EU membership, and administrative consolidation, have been significant issues in rural development. The authors conclude that the development programmes and projects themselves have been insufficient to rescue the Finnish countryside – what is needed are policies and strategies at national and international levels. They also emphasise the point that connections have to be drawn between rural development and the contextual features of individual countries (small towns and vast rural expanses characterise Finland), and broader structural changes across Europe.

With reference to Norway, another sparsely populated Nordic country where regional development is synonymous with rural development, Nina Gunnerud Berg and Hans Kjetil Lysgård assess the effects of processes of industrial and social change on rural development. Norway makes an interesting comparitor because it is not an EU member. The most significant socioeconomic changes in the post-war period – counter-urbanisation and regional economic structuring – are considered as the background within which rural/regional policy has been developed. Economic and social changes provide the context for the development of Norwegian social representations of rurality and the urban. The authors suggest that these social representations relate to processes of regional development, particularly migration flows, in ways far underestimated in Norwegian regional development research and in policies for regional development.

In the concluding chapter, Keith Halfacree, Imre Kovách and Rachel Woodward draw out a number of observations on the broad similarities and differences between the countries under consideration in terms of local political responses to broader processes of rural restructuring. A key conclusion is that given the diversity evident in rural Europe, it is imperative for both policy-makers and academics to draw on the lessons of different states for the development of effective policies for rural development. Furthermore, these need to respect desires for the equitable distribution of resources and for democratic local control and influence over the development process. A further conclusion concerns the nature of international comparative studies such as this, where, we would argue, consideration needs to be given to the ways in which national contexts and specificities can be seen to shape the conceptual approaches taken towards the study of power, politics and rural restructuring across Europe.

References

Bell, M.M. and Lowe, P. (1998), *Regulated Freedoms: The Market and the State, Agriculture and the Environment*, Working Paper 35, Centre for Rural Economy, University of Newcastle upon Tyne.

Boyle, P. and Halfacree, K. (eds) (1998), *Migration into Rural Areas: Theories and Issues*, Wiley, Chichester.

Bruckmeier, K. (2000), 'LEADER in Germany and the Discourse of Autonomous Regional Development', *Sociologia Ruralis*, vol. 40, pp. 219–27.

Buller, H. (2000), 'Re-creating Rural Territories: LEADER in France', *Sociologia Ruralis*, vol. 40, pp. 190–99.

Cloke, P. and Little, J. (eds) (1996), *Contested Countryside Cultures: Otherness, Marginalisation and Rurality*, Routledge, London.

Frouws, J. (1998), 'The Contested Redefinition of the Countryside: An Analysis of Rural Discourses in the Netherlands', *Sociologia Ruralis*, vol. 38, pp. 54–68.

Goverde, H., Cerny, P.G., Haugaard, M. and Lentner, H.H. (eds) (2000), *Power in Contemporary Politics: Theories, Practices, Globalizations*, Sage, London.

Granberg, L. and Kovách, I. (eds) (1998), *Actors on the Changing European Countryside*, Institute for Political Sciences, Budapest.

De Haan, H. and van der Ploeg. J.D. (1994), *Endogenous Rural Development in Europe: Theory, Method and Practice*, Proceedings of a seminar held in Villa Real, European Commission, Luxembourg.

Halfacree, K.H. (1993), 'Locality and Social Representation: Space, Discourse and Alternative Definitions of the Rural', *Journal of Rural Studies*, vol. 9, pp. 23–37.

Halfacree, K.H. (1994), 'The Importance of 'the Rural' in the Constitution of Counterurbanization', *Sociologia Ruralis*, vol. 34, pp. 164–89.

Halfacree, K.H. (1995), Talking about Rurality: Social Representations of the Rural as Expressed by Residents of Six English Parishes', *Journal of Rural Studies*, vol. 11, pp. 1–20.

Hoggart, K., Buller, H. and Black, R. (1995), *Rural Europe: Identity and Change*, Arnold, London.

Kovách, I. (2000), 'LEADER, a New Social Order, and the Central- and East-European Countries', *Sociologia Ruralis*, vol. 40, pp. 181–9.

Marsden, T. (1998), 'New Rural Territories: Regulating the Differential Rural Space', *Journal of Rural Studies*, vol. 14, pp. 107–17.

Marsden, T. (1999), 'The Consumption Countryside and its Regulation', *Sociologia Ruralis*, vol. 39, pp. 501–21.

Murdoch, J. and Marsden, T. (1994), *Reconstructing Rurality: Class, Community and Power in the Development Process*, UCL Press, London.

Osti, G. (2000), 'LEADER and Partnerships: the Case of Italy', *Sociologia Ruralis*, vol. 40, pp. 172–80.

Pratt, A.C. (1996), 'Discourses of Rurality: Loose Talk or Social Struggle?', *Journal of Rural Studies*, vol. 12, pp. 69–78.

Ray, C. (1998), 'Culture, Intellectual Property and Territorial Rural Development', *Sociologia Ruralis*, vol. 38, pp. 1–20.

Ray, C. (1999a), 'Endogenous Development in the Era of Reflexive Modernity', *Journal of Rural Studies*, vol. 15, pp. 257–67.

Ray, C. (1999b), 'Towards a Meta-framework of Endogenous Development: Repertoires, Paths, Democracy and Rights', *Sociologia Ruralis*, vol. 39, pp. 521–37.

Ray, C. (2000), 'The EU LEADER Programme: Rural Development Laboratory', *Sociologia Ruralis*, vol. 40, pp. 163–71.

Shucksmith, M. (2000), 'Endogenous Development, Social Capital and Social Inclusion: Perspectives from LEADER in the UK', *Sociologia Ruralis*, vol. 40, pp. 208–18.

Shucksmith, M. and Chapman, P. (1998), 'Rural Development and Social Exclusion', *Sociologia Ruralis*, vol. 38, pp. 225–42.

Starosta, P., Kovách, I. and Gorlach, K. (eds) (1999), *Rural Societies Under Communism and Beyond: Hungarian and Polish Perspectives*, Lodz University Press, Lodz.

Tovey, H. (1998), 'Rural Actors, Food and the Post-Modern Transition', in L. Granberg and I. Kovách (eds), *Actors on the Changing European Countryside*, Institute for Political Sciences, Budapest, pp. 20–44.

Woods, M. (1997), 'Discourses of Power and Rurality: Local Politics in Somerset in the 20th Century', *Political Geography*, vol. 16, pp. 453–78.

Woods, M. (1998), 'Advocating Rurality? The Repositioning of Rural Local Government', *Journal of Rural Studies*, vol. 14, pp. 13–26.

Chapter Two

Changing Patterns of Leadership and Local Power in Rural Ireland

Chris Curtin and Tony Varley

Introduction

One way of getting to grips with the topic of new patterns of leadership and local power in the contemporary Irish countryside is to try to establish some baseline or baselines against which change can be measured. Such a search would in fact appear to be essential if we are to place some of the change that has been occurring recently in its proper perspective. But before we can turn to the question of baselines we must ask how the three elements of the 'local', 'leadership' and 'power' are to relate to one another at the conceptual level. The distinction between 'power over' and 'power to', conventionally made by students of power (see Goverde et al., 2000, pp. 37–8), can be pressed into service to suggest a relationship between these three elements.

Some mixture of the dimensions of 'power over' and 'power to' are arguably always present in the phenomena of local leadership and power. Two ideal-typical styles of leadership – based on a preponderance either of the 'power over' or 'power to' dimensions – can nonetheless be imagined. Our argument will be that one prominent pattern of rural leadership in the Irish past, and the structure of local power which it at once reflected and reproduced, was heavily based on 'power over' relationships of domination and subordination. What separates such a pattern from a recently emerging pattern of leadership associated with the area-based partnerships is not that the 'power over' dimension has disappeared but that the 'power to' and 'power over' dimensions have come to be related to each other in a very different way.

Our discussion will be arranged in two parts. The first of these involves us in a search for baselines of leadership and local power in pre-1990s rural Ireland. It is in this opening section that we will look at how the 'power to' capacities that characterised certain patterns of local leadership came to be founded on a range of 'power over' relationships of domination and subordination. The second part of our discussion will examine the appearance of an emerging new pattern of leadership and local power that is tied to the proliferation of area-based partnership activity in 1990s Ireland. The assumption of equality that infuses the official rhetoric of 'partnership' and 'participation' surrounding the area-based partnerships might seem to imply that the 'power over' dimension of leadership and local power is becoming insignificant. Yet the emerging pattern of leadership and local power associated with the area-based partnerships has by no means marked a complete break with pre-1990 patterns. Local merchants and politicians, in particular, have not disappeared from the scene. More than anything, the 'power over' dimension has taken on a new guise given the capacity of the European Union (EU) and the Irish state to lay down all the important parameters that guide the organisation of area-based partnership activity.

Pre-1990s Patterns of Leadership and Local Power

Three institutional spheres – the economic, the political and the religious – provide the bases of analytically separate though empirically interconnected arenas of local power and leadership in rural Ireland.

The Economic Sphere: the Gombeenman

For Eipper (1986), as for the earlier account of Gibbon and Higgins (1974) upon which he draws heavily, the economic sphere is accorded primacy. Gibbon and Higgins, drawing on evidence relating to the 'Congested Districts' of the late nineteenth and early twentieth century west of Ireland,[1] paint a picture of local leadership that revolves around the figure of the shopkeeper as an exploitative 'gombeenman'.[2]

The pre-independence gombeenman, as found along the length of the western seaboard, is presented as being capable of using his power over his customers to issue shop credit and provide money loans to create ties of personal dependence. The 'power over' his clients which these transactions conferred allowed the gombeenman to become 'the effectual ruler of large tracks of country and hundreds of "subjects"' (Gibbons and Higgins, 1974, pp. 31–2). Gombeenmen also became adept at using 'ties of monopolistic economic personal dependence to secure a *bloc* of support for electoral and other political purposes' (p. 33). Besides their economic and political power, the gombeenman is presented as being capable of wielding 'ideological patronage'. It was this that allowed him to impose 'an effectual moral order of his own on a locality, usually supported in this venture by a semi-dependent curate or parish priest' (p. 33).

The reign of the gombeenmen attracted some resistance when the Irish cooperative movement decided to set up village banks in the late nineteenth and early twentieth centuries (Bolger, 1977, pp. 156–82). After a promising start, these cooperative credit societies began to falter during the First World War. 'Credit-retailing', Gibbon and Higgins (1974, p. 34) suggest, 'combined with gross indebtedness remain[ed] the rule rather than the exception on the west coast' as late as the early 1970s. Client ties, however, were by then being used to 'gain electoral office for the gombeenman's own personal economic aggrandisement' (Gibbon and Higgins, 1974, p. 35). And with the move towards 'the mobilisation of client ties to secure electoral office', the 'network of dependence among clients' was coming to be based upon 'bureaucratic favours' (p. 35).

Some traces of the gombeenman of old are to be found in Brody's (1974) account of the west of Ireland district identified as 'Inishkillane'.[3] Brody (1974, p. 105) gives an account of a shopkeeper-hotelier applying pressure, with the help of the local priest, to dissuade a man from selling a plot to strangers intent on building a holiday home. He also considers the activities of Michael Ryan, a shopkeeper who has used the capital accumulated in his shop to buy land and become involved in a range of other economic activities. Although 'active in Fianna Fáil',[4] Ryan had not yet become an elected politician (p. 200). It seems that some of Ryan's land was acquired when he used his power to induce an ageing bachelor to sell his farm to him. Compared to the smallholders, content to carry on farming according to a traditional mode, Brody identifies commercial middlemen (such as Ryan) as the dynamic agents of capitalist development in agriculture. Likewise, shopkeepers are seen as benefiting most from the collapse of smallholder society in Eipper's southern district of Bantry in the 1980s. To settle unpaid shop debts, local shopkeepers are claimed to have seized land or purchased it cheaply, and to have subsequently make large financial profits, selling sites when second home tourism began to boom locally in the 1960s (Eipper, 1986, p. 30).

A rather different interpretation of the position of rural shopkeepers is offered by Clark (1979), in his study of the social context out of which the Land League of the 1870s came to mobilise in the west of Ireland as an agrarian/nationalist movement. Clark accepts that the social standing of shopkeepers rose in the post-Famine period. Their ascent was such that they 'came to rival landowners and clergymen as wielders of local power and patronage. They often enjoyed a special relationship with rural people that was comparable even to that of the parish priest' (p. 128). The local standing of rural shopkeepers was based in part on 'the strong kinship ties between this social group and the farming population' (p. 128). The shopkeeper's influential position, however, also reflected the fact that 'he had a much wider circle of friends' than did the farmer; he was more 'cosmopolitan' in the sense that he had 'greater contact with people beyond his own community' (p. 132).

At first glance the picture Clark paints of shopkeepers, reflecting the reality that 'the principal source of credit for small farmers was the small-business class' (Clark, 1979, p. 130), appears to have a clear 'power over' side to it. He certainly acknowledges that the credit system inevitably 'gave rise to conflict between shopkeepers and farmers' (p. 130). Simultaneously, however, he maintains that 'in an effort to maintain their clientele, shopkeepers were known to advance credit for years without payment and even to write debts off if customers could not afford to pay' (pp. 130–31). The implication of these practices is to mitigate the shopkeepers' power over their creditors. Clark is inclined, in the end, to see the relationship between shopkeepers and their customers in terms of mutual dependence and advantage. As he puts it, in spite of the possibility of conflict, 'the credit system contributed to a mutual bond between them. The shopkeeper depended on his customers for business, and the farmer depended on his shopkeepers for goods and for credit' (p. 132).

Further complexity is introduced when we consider Gulliver and Silverman's (1995) historical study of the merchant class of Thomastown, a small rural town located in the southeastern county of Kilkenny. No trace of the gombeenman phenomenon was found here, leading Gulliver and Silverman to lament the practice of generalising from the experience of the west to the rest of Ireland in the setting of ethnographic baselines. Nor were Thomastown shopkeepers involved in supplying leadership to agrarian/nationalist movements (such as the Land League) in the late nineteenth and early twentieth century. The pattern, instead, was one of 'detachment ... from political advocacy and action' (p. 316).

Yet the Thomastown merchants, when organised in their own development association in the 1970s, took for granted and acted as if they had a certain 'power over' the townspeople they claimed to represent. As Gulliver and Silverman (p. 335) put it, their 'status – class gave them the right to define what the important local issues were and to speak not only for their own interests but for what they conceived as the interests of all townspeople'. The formation in 1970 of a development association was crucial here. Gulliver and Silverman (p. 335) suggest that it 'was only after their status-class superiority, and their pragmatic conservatism, were linked to an assumption of leadership through formal organisation, that the intent to establish a retailers' hegemony was openly declared'. In the course of the controversy surrounding the introduction of a one-way traffic system, which they opposed, the Thomastown merchants made explicit the view that their interests were superior to all others:

> TDA [Thomastown Development Association] members expressed an agreed view that 'we *are* the town'. 'We are what has made the town and keep it going' and 'we know best'. Other townspeople – those who favoured the scheme at the Community Council meeting, in talk in pubs and shops, and in private conversations – were 'not the *real* people of Thomastown'. 'They don't know about these things'. 'They don't have the experience'. 'Their opinions are not important' (Gulliver and Silverman, 1995, p. 333, original emphasis).

Although to some extent the merchants' 'hegemonic notions were contested' – we hear how 'many townspeople were suspicious of their motives and resented their assumptions' (p. 335) – the fact that they had an 'organisation to express opinion, to engineer protest, and to secure protection' meant that they could assert themselves as an effective force in the 1980s, 'at least as far as town traffic control was concerned' (p. 335).

The Political Sphere: Clientelism and Subordination

We turn now to the question of leadership and local power as it involves politicians in the sphere of local politics. We have already seen from some of the accounts (Gibbon and Higgins, and Clark, especially) that shopkeepers were able to convert their economic power so as to develop careers as local or even national politicians. This pattern was to long persist. As late as the 1960s, small-scale merchants and farmers formed 'the solid core of the Dáil [parliament] and comprise[d] more than half of its total membership' (Chubb, 1971, p. 211). Dáil deputies, Chubb (p. 211) observes, 'are particularly active in local government, which is one of the routes to parliamentary office and which offers the most frequent opportunities for service to their constituents'.

There is a literature that sees the servicing of constituents as giving rise to a form of clientelism that politicians must heavily invest in if they are to hold their seats. By this reckoning, one's competence as a clientelist politician depends on perceptions of how much to deploy 'power to' to deliver benefits to one's constituents. To the extent that those constituents who benefit from favours feel morally obliged to repay their benefactor come polling day, politicians might be said to have a certain 'power over' their clients. The real power they had over their clients, however, was rooted in the reality of popular subordination. One prominent contention in the literature is that clientelism, far from lessening or extinguishing this subordination, actually operated to reproduce and deepen it.

Clientelism and subordination appeared mutually reinforcing. On the one hand, both Sacks (1976) and Higgins (1982) note the tendency for clientelism to discourage the formation of self-help groups among the subordinate. These writers blame clientelism for individualising and privatising social problems, thereby discouraging collective action by disadvantaged elements and perpetuating them in their position of economic and social inequality. On the other hand, under both colonialism and independence, the distrust that weaker elements of rural society had of state bureaucracy, and their belief that bureaucratic doors would open only when pressure was applied by some powerful broker, contributed to clientelism's persistence.

Centralism and bureaucratic control became ever more deeply embedded as institutional tendencies within the state-building effort in the early decades of political independence.[5] 'After independence', Chubb (1971, p. 279) notes, 'the trend towards grassroots authorities was reversed, a process that has continued ever since.' The poor law unions were dissolved in 1923; in 1925, it was to be the turn of the rural district councils, whose powers were transferred to the county councils. Subsequently, the creation of a manager-based system of local government administration ensured that local government became steadily more bureaucratised, though by the 1940s this was to provoke 'a conscious attempt to turn back the tide of managerial bureaucracy and to give the elected members the last say' (p. 286).

The authoritarian tendencies of Irish society have been seen to at once reflect and underpin the sense of subordination that pervades clientelist politics in Ireland. As evidence of 'Irish people's deference to authority', Coakley (1996, p. 37) reminds us how the commentators on Irish politics have been struck by 'the high degree of public acquiescence in decisions by governments to postpone local elections, or even to suspend local councils and replace them by appointed commissioners'. Social surveys of the late 1960s and early 1970s documented the prevalence of the popular view that the welfare of the country depended significantly on the availability of 'a few strong leaders' or 'a good strong leader' (p. 37). The 'pressure for conformity' has been judged to have been especially strong in rural areas (Gallagher, 1982, p. 19).

Clientelism has been claimed not only to operate against the interests of the subordinate but to the benefit of economically advantaged groups. How clientelism has contributed to the growing power of the bourgeoisie is a theme

Eipper (1986) in particular pursues. Most of his attention focuses on Fianna Fáil and its ability to combine clientelist populism with policies tailored to the needs of the bourgeoisie. The party's success in attracting cross-class support is attributed to its clientelist tendencies that 'co-opted the subordinate classes' and that 'helped mediate divergent class interests', without 'jeopardising the expansion of the indigenous bourgeoisie' (pp. 84, 65, 66). All this was to allow the real business of running the state primarily in the interests of the dominant class (the indigenous bourgeoisie and, increasingly as time went by, foreign capital) to proceed relatively smoothly (pp. 84–5). Likewise Hazelkorn (1986, p. 327) sees clientelism operating to ensure that 'inter-class linkages are stressed, while intra-class solidarities, which stress (political) action via class, are de-emphasised'. Clientelism in Ireland is seen to have 'been reinforced and reproduced by the state, resulting in the masking and deflecting of conflict'. The result of this is that 'the state can be experienced as representing society's "general interest"' and that 'class issues are marginalised and seen as inappropriate' (p. 333).

The Religious Sphere: the Catholic Church

To what degree were pre-1990s patterns of leadership and local power grounded in the institutional sphere of religion? By the close of the nineteenth century the Catholic Church in Ireland had built a formidable ideological and institutional presence for itself. This had been achieved on the strength of internal reform and reorganisation under tight central control. For example, the church projected its institutional and ideological presence through the deployment of a large contingent of clergy who staffed the local parishes and frequently became active in many areas of public life. In addition, a 'devotional revolution' had resulted in 90 per cent Sunday mass attendance by the 1870s (Hynes, 1978, p. 137).

From the second half of the nineteenth century, Catholicism's appeal has been linked to how well its message and practices appealed to certain classes and categories, such as the rising 'commercially oriented farmer class' (Hynes, 1978, p. 147) and 'women of the house' (Inglis, 1998). The church's message to the socially subordinate elements of Irish society was to submit themselves to external authority and to discipline their wants and impulses, especially in matters relating to sexuality. Catholic moral teaching put down deep roots. Yet, clerical power ran up against definite limits in its ability to intervene in certain facets of social life. Late marriage and permanent celibacy, emigration, and agrarian and political violence were to attract much clerical unease and fulmination since, as Hynes (1978, pp. 147–8) points out, clerical views on these matters were simply ignored as a rule.

Although the church's castigation of physical violence drew it into controversy during the independence struggle (1916–21) and civil war (1922–23) that preceded native rule, it was to suffer no long-term damage as a result. Writing of the period when much of the revolutionary elite had fallen foul of the Church authorities for the manner they had resorted to physical force, Thornley points out how:

> The shock of first revolution and then civil war, the formal if temporary censure of half the country's political leaders, was somehow absorbed by the enormous resilience of the folk church. When the dust had settled, the parish priests were scarcely less the leaders of their communities than they had been in pre-Fenian times. Indeed, the vacuum left by the old aristocracy drew them, if anything, still closer to the centre of community decision-making. Their social and cultural eminence automatically brought them to the chairmanship of cultural, athletic and economic community organisations. The British government had historically recognised their status by tacitly incorporating them in the local management of its national education system. The twentieth century still found them playing a significant part in social and political life (cited in Chubb, 1971, p. 54).

Indeed, several commentators suggest that the Catholic Church's grip on Irish civil society was to tighten even more in the decades after the creation of the Irish Free State in 1922–23. At the very least, the post-independence state, while not a theocracy in any strict sense, was heavily influenced in its social policy by Catholic moral teaching. The outlawing of divorce, the ban on artificial contraception and the imposition of censorship became cornerstones of an approach that saw the state legislating in a way that was mindful of the requirements of Catholic moral teaching.

At the local level, priests were central figures in the public life of their parishes. Writing in the 1960s, one well-placed commentator's assessment was that:

> almost nothing is undertaken, from drama groups to tidy towns competitions, without it being under the chairmanship or presidency of a parish priest or one of his curates ... the view had grown up quite strongly that no activity locally was proper or right unless it had the approval of the local clergy (McCarthy, 1968, p. 110).

Eipper's (1986, p. 104) account of the Bantry district in the 1980s yields insight into the situation of a local 'activist priest' when he observes how:

coming in as an outsider, usually without kinship or other local connections, an activist priest had considerable room for manoeuvre. As a professional 'doer' without prior allegiances to local factions, the new priest (particularly if he brought a reputation with him) was looked upon as a man who could 'get things done'.

Catholic clergy also provided the leadership for some of the national and regional movements that emerged in the pre-1990s period to defend rural interests against an assortment of threatening forces. The still extant parish council movement known as *Muintir na Tíre* (People of the Land), initiated in the 1930s by the Tipperary priest, Fr John Hayes, appealed most to the Catholic middle class. Local notables, priests and schoolteachers in particular, tended to occupy the leadership slots in the local parish councils.

Built around the leadership of the energetic west of Ireland priest, Fr. James McDyer, the 'Save the West' (STW) campaign of the 1960s was inspired by an analysis which suggested that self-help efforts and immediate and effective official action were required to halt the terminal decline of western smallholder communities. Fr McDyer was able to use his abundant clerical connections to mobilise popular support for the STW campaign. The local business notables – organised in chambers of commerce and development associations – were especially active in organising the early meetings that led to the emergence of the STW campaign in Mayo (Varley and Curtin, 1999).

A 'Ruling Trinity'?

The Save the West campaign shows how clerical and business elites could work together. Eipper (1986), on the basis of his Bantry fieldwork, goes further by positing the existence of a 'ruling trinity'. This involved clergy, local business people and politicians pulling together in pursuit of projects of clear local importance but which Eipper sees as serving the interests of the bourgeoisie above all. From the 1950s, for instance, under the stimulus of the 1952 Undeveloped Areas Act which provided for regional development incentives to attract manufacturing industry to the west, local business leaders, politicians and clergy began campaigning to attract industrialists to Bantry, offering land, cheap labour and local goodwill as the principal incentives. The 'ruling trinity', organised in the Bantry Development Association, achieved its greatest coup when Gulf Oil decided to build a large oil storage facility in Bantry Bay in the mid-1960s.

What conclusions can be drawn about patterns of leadership and local power from our discussion so far? We began with the suggestion that the 'power to' represent and organise local interests accumulated by shopkeepers, politicians and priests in the rural Ireland of the pre-1990s period was heavily based on a range of 'power over' relationships of domination. Exploring this suggestion has involved us in a search for baselines. Nowhere is the complicated nature of this search more apparent than in the case of shopkeepers. Differences between accounts of merchant power can be related to the interpretative schemes different writers have relied on as well as to actual differences between places in the west, east and south of Ireland. Enough has nonetheless been said to demonstrate the manner in which the leadership slots available in local society had come to be occupied by merchants, clergy and clientelist politicians. What is equally clear is that the power of these elements was based on various 'power over' relationships of domination and subordination.

Patterns of Leadership and Local Power in the 1990s

How have the 'ruling trinity' fared in the 1990s in the face of the rise of the 'area partnership'? Over the past decade or so, 'local communities', as one element of a new local development strategy revolving around area-based partnerships, have been invoked in official circles as holding part of the solution to some of Ireland's economic and social problems. Local communities have been seen in official eyes to possess a capacity to mobilise local support, engage in enterprise and create employment as well as contribute to the delivery of a range of social services. The most striking feature of this official attribution of such capacities to 'local communities' is that historically it represents a new departure. This is evident when we remember how a heavily centralised Irish state came to deal with local community interests in the pre-1990s period.

The Rise of Area Partnerships

The early view taken by the new Irish state of 'local communities' was that they were largely irrelevant to the accumulation process and to social service provision.[6] What nationalist commitment there was to reorganise the economy using local cooperatives during the Anglo-Irish War (1919–21) was never more than partial (O'Connor, 1988, p. 93). In the 1920s, the first Minister for Agriculture resisted pressure to make 'a public declaration in favour of co-operation' (Bolger, 1977, p. 118). Nevertheless, the state in Ireland did come to see local parish councils as a resource to be mobilised as part of the national response to the wartime emergency during the Second World War. Furthermore, some few state officials, as the state hesitantly committed itself to a regional industrial policy in the 1950s, did see community groups playing an indirect role in hosting manufacturing industry and making it feel welcome in the localities where it chose to locate (Curtin and Varley, 1986). From the mid-1960s within the Gaeltacht (Irish-speaking) areas, community-based cooperatives had begun to be officially encouraged.

We have to await the late 1980s and the advent of the area-based partnerships, however, for the real breakthrough in the state's estimation of the development capacities of the 'local community'. At this point the state's interest in 'communities' as actors in local development began to break new ground. We thus find the National Development Plan 1989–93 (Government of Ireland, 1989, p. 65), in addressing the topic of integrated rural development, announcing that 'the priorities will be set by the communities themselves, who will be responsible for ensuring that local potential will be realised'. This document's successor, the National Development Plan 1994–99 (Government of Ireland, 1994, p. 69), speaks about 'empowering communities to sponsor innovative projects'. The notion of the local community as collective actor was to feature equally centrally in the third European anti-poverty programme (Poverty 3) and in the generation of new area-based rural development programmes – such as LEADER I and II and INTERREG – emanating from the European Economic Community/Union (EC/EU). A leading theme for the area-based partnerships has been the combating of social exclusion (conceived in terms of multiple disadvantage).

Two interpretations of the area partnerships in Ireland can be introduced here. For the optimists, the official rhetoric of transferring power to the local community *via* the area partnerships is to be taken seriously. Participation can pave the way for the 'empowerment' of local community interests. The pessimistic interpretation, on the other hand, makes two points. It suggests, firstly, that community actors are likely to be little better than puppets as long as state or EC/EU elites, by virtue of their strategic location and their vastly superior resources, are able to use their powerful position to dominate the area partnerships. A second pessimistic point, reflecting the operation of local power structures, is that the traditional notables of local society are likely to take control of community interests within the arena of the area-based partnerships. Taken together, domination by official and local elites leads to the conclusion that the prospects for popular empowerment *via* the local area partnerships can only be poor.

In the light of these two contrasting interpretations, three questions, bearing on the possibility of new patterns of local leadership and power, can now be posed. Are the area partnerships controlled by the state/EU 'official' elites? Do 'traditional' local notables, the mercantile-politico-clerical 'ruling trinity' discussed above, control them? Or is it that they are controlled by a new strand of local leadership that has come to the fore along with the area partnerships? Our discussion will draw on three of the more well known of the rural area partnerships. These are the LEADER I and II partnership programmes; the Programme for Economic and Social Progress (PESP) Pilot Initiative on Long-term Unemployment (1991–93) area partnerships that were the prelude to the Area Development Management (ADM) partnerships; and the Forum partnership (1990–94) of northwest Connemara, the only rural project of the EU's Poverty 3 programme to be sited in the Irish Republic. Both the area partnerships themselves, and the supra-local alliances to which they have given rise, will be considered.

Control by Official Elites

The official rhetoric behind the new area-based initiatives clearly accepted that neither the state nor the EC/EU could supply all the answers to local development problems. The time had come for local actors (including organised local communities) to draw on their knowledge and resources and avail of the opportunity to become more 'central players' in the sphere of local development. The pessimistic view would be that all this conceals the way official control is used to steer the local partnerships. How, our pessimist might ask, can local and community interests hope to play a shaping role when

> the basic parameters of [these] development programmes – the guidelines concerning substantive content, the areas to be included, the size of these, the respective roles of the 'partners', the money (if any) to be expended/subscribed, and the timetables – have all been decided in advance (Varley, 1991, p. 106).

Certainly, at the start-up phase of the area partnerships, the evidence for official control is very strong. Rather than the product of local demands or the mobilisation of local pressure, the origins of each of our three area partnerships in 1990s Ireland have to be located within a process of top-down decision-making. In fact, if we leave the PESP programme aside, most of the thinking behind the area partnerships originated more in the EC/EU than in the Irish state. Does all this mean that local community interests were not involved at all? Sometimes local community groups – Ballyhoura in east Limerick in LEADER I and Connemara West in Poverty 3 are good examples – made the running in assembling partnership groups and preparing applications. The final decision about the selection of local partnership sites, however, was left in the hands of the Irish state and the EC Commission.

How central was the contribution of community interests to the actual operation of the partnerships? In sharp contrast to the official rhetoric that took the 'local community' as possessing considerable power, the reality that quickly confronted the organisers of many of the area partnerships was one in which the community as a collective actor was weakly organised relative to other partnership strands. To take the PESP area-based initiative, for example, a number of difficulties were to arise around the identification of the community constituency and the selection of community representatives in a manner that would satisfy the requirements of democratic accountability (see Craig 1994, p. 20, pp. 31–2; Combat Poverty Agency, 1995, p. 4, pp. 9–24).

There is some evidence from LEADER I to suggest that the community partners tended to be eclipsed by their state partner colleagues. As the official evaluators observe (Kearney et al., 1994, p. 110), the statutory partners found that their administrative skills and the control of resources left them in a superior position. 'The community representatives', the evaluators further note, 'are sometimes not sufficiently skilled to enable them to participate effectively in a formal Board, especially when confronted by more skilful representatives of the statutory sector on whom they depend for many resources' (p. 110). Yet, from another perspective, the evaluators suggest on the same page that 'the inflexibility of agency representatives was a cause of deep frustration for other partners'. Writing of the PESP partnerships, Craig (1994, p. 63) similarly points out that 'power structures at local level have not changed. Statutory agencies still maintain control over resources and there has been considerable resistance to any proposal that might alter this' (see also Webster, 1991).

In the light of experience, community interests have come to be seen in official eyes as standing to benefit by participation in the partnerships only if they are prepared to adopt certain organisational disciplines. To be taken seriously as 'partnership' material, community interests had little option but to adopt a more systematic, professionalised approach that entailed moving towards a full-time commitment and a heavier dependence on paid staff.

For example, it was realised early on in Forum that local community interests would have to undergo significant organisational 'capacity building' if they were to contribute more fully to local development. For this reason a full-time community development worker was assigned to work with local community and community-based groups. What is clearly significant is that many community activists would concur with the state's view that the local community's contribution can be improved if a more rational mode of organisation is adopted. Such concurrence has arguably improved the position of the paid staff vis-à-vis the volunteers among organised community interests as well as throughout the area-based partnerships more generally.

A hardening of the organisational structure was also apparent for PESP. General responsibility for the direction and supervision of the PESP area partnerships rested with a central review committee. This convened an operational team made up of representatives of employers, trade unionists, FÁS (the National Training Authority) and the Department of Education. A formal distancing of the area-based initiative from the Irish state proved necessary with the appearance of the EC's global grant in 1992. It was at this juncture that Area Development Management (ADM) Ltd. was formed. Hard upon ADM's appearance the broader community development dimension to the PESP partnerships began to be taken more seriously. Funding, under the provisions of the EC's global grant, was now allocated to 'capacity building among local organisations' (Craig, 1994, p. 95).[7]

Control by Traditional Elites

Have our 'traditional' local notables in the pre-1990s 'ruling trinity' enjoyed any success in dominating the area partnerships? The picture is mixed. While Catholic priests have been conspicuous by their absence for the area-based partnerships,[8] the same cannot be said of the local business class. A glimpse of the power of local business elites is supplied by O'Reilly's (1994, p. 137) in-depth study of one of the 12 PESP partnerships. What this shows is the ease with which commercial interests were able to extend their strength in local society into the area partnerships. For example, nine of the 12 PESP partnership board chairpersons appear to have had business backgrounds (Craig, 1994, p. 22). The power of local commercial elites in some of the LEADER I partnerships is hinted at by the Comptroller and Auditor General when it is noted that 'some of the projects grant aided were promoted by Board members or by persons with whom they were closely connected'. The Comptroller and Auditor General (Office of the Comptroller and Auditor General, 1995, p. 17) observes that this gives '... rise to a possible conflict of interest'. In response to such criticism, the state took steps to ensure that the LEADER II area partnerships (1995–99) would be subjected to a framework of much tighter central control (see *Irish Times*, 27 September 1994).

Local government politicians have always regarded themselves to be the most legitimately constituted representatives of local communities in Ireland (see Ó Cearbhaill and Varley, 1988). It is for this reason that many of them have resented what they see to be their own exclusion and the privileging of community actors under the area partnerships (see also Sabel, 1996, pp. 85–6; Walsh et al., 1998, p. 218). The grievance of these local politicians has been that they, as duly elected public representatives, have a superior right to all others, regardless of how 'representative' these others may claim to be, to represent local communities. This contention appeared to have been substantially accepted in the 1996 proposals for local government reform. As well as calling for a restructuring of the administrative framework of the Irish area partnerships by concentrating overall responsibility in the hands of the Department of the Environment and Local Government, the proposals moved to make room for local authority representatives. A White Paper (Department of the Environment, 1996, p. 27) announced that 'the Government has decided that the local government and local development systems will be integrated on completion of the current round of structural funds programmes'.

Steps towards the realisation of such an aim have been set forth in the *Task Force on Integration of Local Government and Local Development Systems* report. One of the Task Force's recommendations was that 'Local development groups set up under EU programmes, be they Partnerships, LEADER groups, or other such groups, shall post-1999 have county/city councillors on their boards' (Department of the Environment, 1998, p. 14). Simultaneously, the Task Force accepted that the 'community and voluntary sector' be given representation on the new County/City Development Boards (p. 15). Fearing that all this might lead to a take-over and the intrusion of a clientelist culture and practices, some of the ADM partnerships have expressed their apprehensions (European Social Fund Evaluation Unit, 1999, pp. 153–4). Since their establishment, the recently established County and City Development Boards have brought together representatives from local government, local development organisations, state agencies, social partners (such as organised employers and trade unions) and community and voluntary associations. The boards, on which the representatives of community and voluntary associations are significantly outnumbered, have been charged with the strategic planning of economic, social and cultural development and with ensuring the adoption of an integrated approach to implementing the measures identified in the plan.

To coordinate the provision of services and to monitor the implementation of programmes arising from EU Structural and Cohesion Funds, eight regional authorities, comprising councillors nominated by the local authorities, were established in 1994. As the whole country was then designated an Objective 1

Region, central government was the regional authority for the management of EU Structural Funds in the 1994–99 period. Over the current phase of Structural funds (2000–06), however, Ireland will have two designated regions – the Border, Midland and Western (BMW) Region, eligible for Objective 1 levels of funding, and a Southern and Eastern Region covering the remainder of the country (see Government of Ireland, 1999). Two new regional assemblies have been established to set detailed priorities and oversee the implementation process. Their memberships are drawn from elected representatives of the relevant regional authorities. Taken with the appearance of the County and City Development Boards, these developments indicate that politicians have successfully fought back to reclaim a good deal of the ground they believed they had lost with the advent of the area partnership schemes in the early 1990s.

Control by New Community Leaders

How well have community activists succeeded in exerting control over the area partnerships? Again, the picture is mixed. In spite of the professed 'community' character of the PESP partnerships, there is evidence to indicate that the representatives of community interests came to view themselves as the weakest link in the partnership chain. Numerical equality in the area partnership directorates did not clear the way for participation on anything like equal terms. As was also true of LEADER I, the prior experience of 'community directors' seems not to have prepared them for participation in the PESP pilot initiative. The slow, consensual decision-making style favoured by some community interests and the more executive, decision-making style of leadership that applied to private sector and state administration were always in danger of clashing (Craig, 1994, p. 90). On occasion, community actors have also found themselves in a disadvantaged position in their relations with the paid staff of the area partnerships. It seems that the technocratic language used by the PESP area partnership professional staff contributed to feelings of marginalisation among community directors (O'Reilly, 1994, p. 189).

On the other hand, some community activists were allowed or acquired considerable discretion over such matters as the composition of boards and committees, the setting of agendas and the prioritising and re-prioritising of project activity. Sometimes, as in LEADER I, 'community partners' found themselves in a position where they could steer the work of the partnerships in a certain direction (Kearney et al., 1994, p. 109). Forum's adoption of a 'learning process' approach, which sought to systematically take account of the lessons of experience, resulted in some significant changes being made to the composition of the project's board and to its programme of work. A noteworthy feature of these changes was that they embodied the wishes of community actors as much as those of state partners.

An important stimulus for change in Forum was the struggle (particularly evident among some of the partnership's paid community workers) which developed between the adherents of two contrasting styles of doing community work. In the first of these styles, the emphasis falls on the process of building up the capacity and awareness of local actors in line with participatory and emancipatory ideals. Those who favour this approach take it that community action is naturally a slow, drawn-out process that cannot be speeded up excessively if it is to be done properly. Demonstrating effectiveness in the near term, as expressed in the ability to deliver some tangible and measurable 'end product' within a set time period, is what gives the second way of doing community work its distinctive style (see Lumb, 1990). The tension between these two differing styles, one that ran through all of the local partnerships on account of their limited life-spans to some extent, was to surface especially forcefully in Forum.

What this tension also illustrates is how a new strand of local leadership is taking shape around the pressures and the opportunities that have accompanied the area-based partnership schemes. The appearance of typically credentialised development workers, as a new strand of local leadership, is even more marked in the supra-local networks or alliances that have accompanied the various area-based partnership programmes. Thus, the leading lights in the ADM and LEADER networks have been the managers and chairpersons of the individual partnership groups. These networks have acted, in their defensive mode, to ensure that local partnerships are not deprived of funding or shut down. The more positive hope in forming supra-local alliances has been that these networks can put local partnerships in a better position to take an overall view and to influence the direction of local policy as well as the fate of the partnership approach at local level. A point made by the LEADER network, for instance, is that there should be an official commitment to fund a cycle of local development that endures through various stages, perhaps for as long as 15–20 years (LEADER, 1999, p. 18). By the same token, Planet (the network of the ADM partnerships) has assembled a number of subcommittees which have produced documents seeking to open up new fronts for local development in the social economy and in the provision of early childhood care and education services. The various networks are represented on a large number of consultative and policy-making national bodies.

Conclusion

At one level, the 'power over' the local partnerships of 1990s Ireland wielded by their official sponsors is obviously formidable. Operating within a framework of officially imposed constraints, the local area-based partnerships and their supra-local networks have nonetheless accumulated sufficient power to make some significant differences. Many of the local actors involved would accept that what has been achieved so far is but a beginning. They would concede nonetheless that a significant start has been made in turning the rhetoric of local empowerment via the partnership approach into reality.

Compared to the 'power over' domination exerted by the traditional ruling trinity, two very relevant features of the emerging pattern of local leadership associated with the area partnerships are professionalisation and official commitment (most visible in ADM and in Poverty 3) to combat social exclusion. The official stress on professionalisation implies that local community actors become more rationally organised collective actors. The emergence of a new strand of local leadership from among the professionalised paid staff, observable in the local partnerships as well as in the supra-local networks, has been deeply influenced by the official desire for greater rationalisation. A major attraction of the supra-local networks in official eyes is that they can be used to help place the organisation of the local partnerships (and the organisation of community interests within the context of the partnerships) on a more rational footing.

The emphasis on social exclusion has encouraged the area partnerships' paid staff to cultivate an interest in 'participatory democracy' and in adopting a community development (i.e. collective) approach to the solving of social problems. These stances imply a desire to escape the 'power over' domination exerted by traditional elites. More particularly, they imply a rejection of clientelism's individualising tendencies and the ruling trinity's bias in favour of serving the interests of the bourgeoisie. Yet, both the local partnerships and the networks have been more interested in establishing their own sphere of 'local development' than in crossing swords openly with the traditional notables. Merchants, in fact, have found a new place for themselves in the partnerships in some instances. The same cannot be said of the Catholic clergy. Over the past decade the Church's 'moral monopoly' has weakened considerably, consequent upon the institutional and morale crisis that has followed the sharp drop-off in vocations, priests exiting the ministry in numbers and a spate of legitimacy-damaging sex and child abuse scandals (see Inglis, 1998; Donnelly, 2000).

Our conclusion is that the institutional framework supplied by the area partnerships has cleared the way for a new strand of professionalised leadership to appear at local level. The emerging pattern is evident within the local partnerships themselves, their associated networks as well as among the community interests with whom they work. The area partnership-based approach, of course, is still evolving and it is impossible to be certain at this stage whether it has a long-term future ahead of it or whether the appearance of politicians will radically change its complexion. What is reasonably clear is that the vital conditions of its development will continue to depend on the commitment of its official sponsors.

Notes

1 The agriculturally disadvantaged areas, officially designated 'congested' under the 1909 Land Act, extended across most of the west of Ireland.
2 'Gombeen', Gibbons and Higgins (1974, p. 31) suggest, 'is a corruption of the Irish word (*gaimbín*) for interest ... the term was originally coined to describe the activities of rural moneylenders in pre-Famine Ireland'.
3 'Inishkillane' is a composite of a number of places that Brody visited during his stay in the west of Ireland in the late 1960s and early 1970s.
4 Fianna Fáil, the major nationalist party in twentieth-century Ireland, was founded by Éamon de Valera in 1926 around the defeated faction in the Irish civil war (1922–23).
5 Under the provisions of the 1921 Treaty, a self-governing Irish Free State (26 counties) came into being within the British Commonwealth. A fully politically independent Republic of Ireland was declared in 1949.
6 The same cannot be said of the voluntary sector more generally, as the case of church-controlled education makes plain (Breen et al., 1990, p. 33).
7 A total of 30 per cent of European Social Fund spending on the Irish area partnerships, as funnelled in 1997 through ADM, went on 'community development actions' (European Social Fund Evaluation Unit, 1999, p. xvi).
8 A prominent exception here is Fr. Michael Mernagh, who has been active in LEADER at the local and supra-local levels.

References

Bolger, P. (1977), *The Irish Co-operative Movement: Its History and Development*, Institute of Public Administration, Dublin.
Breen, R., Hannan, D., Rottman, D. and Whelan, C. (1990), *Understanding Contemporary Ireland: State, Class and Development in The Republic of Ireland*, Gill and Macmillan, Dublin.
Brody, H. (1974), *Inishkillane: Change and Decline in the West of Ireland*, Penguin, London.

Chubb, B. (1971), *The Government and Politics of Ireland*, Oxford University Press, Oxford.

Clark, S. (1979), *Social Origins of the Irish Land War*, Princeton University Press, Princeton.

Coakley, J. (1996), 'Society and Political Culture', in J. Coakley and M. Gallagher (eds), *Politics in the Republic of Ireland*, PSAI Press, Limerick, pp. 25–48.

Combat Poverty Agency (1995), *Community Participation: a Handbook for Individuals and Groups in Local Development Partnerships*, Combat Poverty Agency, Dublin.

Craig, S. (with McKeown, K.) (1994), *Progress Through Partnership: Final Evaluation Report on the PESP Pilot Initiative on Long-term Unemployment*, Combat Poverty Agency, Dublin.

Curtin, C. and Varley, A. (1986), 'Bringing Industry to a Small Town in the West of Ireland', *Sociologia Ruralis*, vol. 26, pp. 170–85.

Department of the Environment (1996), *Better Local Government: a Programme for Change*, The Stationery Office, Dublin.

Department of the Environment (1998), *Task Force on Integration of Local Government and Local Development Systems*, Department of the Environment, Dublin.

Donnelly, J. (2000), 'A Church in Crisis: The Irish Catholic Church Today', *History Ireland*, vol. 8, pp. 12–17.

Eipper, C. (1986), *The Ruling Trinity: A Community Study of Church, State and Business in Ireland*, Gower, Aldershot.

European Social Fund Evaluation Unit (1999), *ESF and the Local Urban and Rural Development Operational Programme*, European Social Fund Evaluation Unit, Dublin.

Gallagher, M. (1982), *The Irish Labour Party in Transition 1957–82*, Manchester University Press, Manchester.

Gibbon, P. and Higgins, M. (1974), 'Patronage, Tradition and Modernization: The Case of the Irish Gombeenman', *Economic and Social Review*, vol. 6, pp. 27–44.

Goverde, H., Cerny, P., Haugaard, M. and Lentner, H. (eds) (2000), *Power in Contemporary Politics: Theories, Practices, Globalizations*, Sage, London.

Government of Ireland (1989), *National Development Plan, 1989–1993*, Stationery Office, Dublin.

Government of Ireland (1994), *National Development Plan, 1994–1999*, Stationery Office, Dublin.

Government of Ireland (1999), *Ireland: National Development Plan 2000–2006*, Stationery Office, Dublin.

Gulliver, P. and Silverman, M. (1995), *Merchants and Shopkeepers: An Historical Anthropology of an Irish Market Town, 1200–1991*, University of Toronto Press, Toronto.

Hazelkorn, E. (1986), 'Class, Clientelism and the Political Process in the Republic of Ireland', in P. Clancy, S. Drudy, K. Lynch and L. O'Dowd (eds), *Ireland: A Sociological Profile*, Institute of Public Administration, Dublin, pp. 326–43.

Higgins, M. (1982), 'The Limits of Clientelism: Towards an Assessment of Irish Politics', in C. Clapham (ed.), *Private Patronage and Public Power: Political Clientelism in the Modern State*, Frances Pinter, London, pp. 114–41.

Hynes, E. (1978), 'The Great Hunger and Irish Catholicism', *Societas*, vol. 8, pp. 137–56.

Inglis, T. (1998), *Moral Monopoly: The Rise and Fall of the Catholic Church in Modern Ireland*, University College Dublin Press, Dublin.

Kearney, B., Boyle, G. and Walsh, J. (1994), *EU LEADER I Initiative in Ireland: Evaluation and Recommendations*, Department of Agriculture, Food and Forestry, Dublin.

LEADER (1999), *Building on Experience*, Discussion Paper prepared by Comhar LEADER na hÉireann in Consultation with Northern Ireland LEADER Network, Comhar LEADER na hEireann, Fermoy, Co. Cork.

Lumb, R. (1990), 'Rural Community Development: Process Versus Product', in H. Buller and S. Wright (eds), *Rural Development: Problems and Practices*, Avebury, Aldershot, pp. 177–90.

McCarthy, C. (1968), *The Distasteful Challenge*, Institute of Public Administration, Dublin.

Ó Cearbhaill, D. and Varley, T. (1988), 'Community Group/State Relationships: The Case of West of Ireland Community Councils', in R. Byron (ed.), *Public Policy and the Periphery: Problems and Prospects in Marginal Regions*, Association for the Study of Marginal Regions, Belfast, pp. 279–96.

O'Connor, E. (1988), *Syndicalism in Ireland 1917–1923*, Cork University Press, Cork.

O'Reilly, P. (1994), *Whose Plan is it Anyway?: Struggles over the Attribution of Meaning in the Planning Process*, MA thesis, Department of Political Science and Sociology, University College Galway.

Office of the Comptroller and Auditor General (1995), *The LEADER Programme*, Stationery Office, Dublin.

Sabel, C. (1996), *Ireland: Local Partnerships and Social Innovation*, OECD, Paris.

Sacks, P. (1976), *The Donegal Mafia: An Irish Political Machine*, Yale University Press, New Haven.

Varley, T. (1991), 'Power to the People? Community Groups and Rural Revival in Ireland', in B. Reynolds and S. Healy (eds), *Rural Development Policy: What Future for Rural Ireland?*, Conference of Major Religious Superiors, Dublin, pp. 83–110.

Varley, T. and Curtin, C. (1999), 'Defending Rural Interests Against Nationalists in 20th Century Ireland: A Tale of Three Movements', in J. Davis (ed.), *Rural Change in Ireland*, The Institute of Irish Studies, Queen's University Belfast, pp. 58–83.

Walsh, J., Craig, S. and McCafferty, D. (1998), *Local Partnership for Social Inclusion?*, Oak Tree Press in association with Combat Poverty Agency, Dublin.

Webster, C. (1991), *How Far Can You Go? The Role of the Statutory Sector in Developing Community Partnerships*, MA thesis, Department of Political Science and Sociology, University College Galway.

Chapter Three

The Politics of Rural Development in The Netherlands

Henri Goverde and Henk de Haan

Introduction

The Netherlands is not particularly associated with rurality. It is one of the most densely populated and urbanised areas of the world, and its position as a leading exporter of agricultural and horticultural products is based on intensive farming in an agro-industrial landscape. In 1997 only 8 per cent of the population lived in low-density, dispersed settlements, and according to the OECD classification the Netherlands has no predominantly rural areas. Most of the Netherlands is part of the 'European Delta Metropolis', a triangular area between Lille, Amsterdam and the Ruhr area, where population densities are higher than 100 inhabitants/km^2. The multiple centres of this Metropolis are connected by infrastructure corridors – a dense network of motorways and railway lines. With some imagination, the intermediate rural areas could be seen as a collection of urban parks and city greens.

However, if we consider the fact that only about 13 per cent of Dutch territory is used for buildings, industry and infrastructure, it is clear that there is a great divide between an overwhelming amount of 'open' or rural space on the one hand, and concentrated urban or semi-urban settlements on the other. In the experience of Dutch people the countryside comprises all areas beyond the built-up environment, even though these areas are becoming smaller and one can never escape from traffic noise and the view of buildings at the horizon. The feeling of being in the countryside is strengthened by the fact that 'open space' in the Netherlands mainly consists of cultivated agricultural land, which means that one is immediately confronted with fields, cows, farms and villages outside the urban environment. Although rural areas are increasingly used for recreational purposes, this is not associated with using urban space, but clearly with 'going out into the country'.

The present concentration of the Dutch population in mainly urban areas and the domination of cultivated land over natural areas (only 14 per cent) are the results of two factors. While the territorial expansion of agriculture came to a halt in most European areas in the nineteenth century, it continued in the Netherlands until the Second World War. Especially in the sandy-soiled and peat areas almost all wasteland was reclaimed, resulting in a 20 per cent increase of cultivated land between 1900 and 1950. During the same period, the population doubled from 5 to 10 million people. After the Second World War, the proportion of agricultural land only diminished slowly from 76 per cent in 1950 to the present 70 per cent. As a result of a relatively high birth, rate the population increased from 10 million in 1950 to almost 16 million at the end of the 1990s. Pre-war population growth was mostly channelled by compact urban growth on the one hand and the expansion of rural settlements. Although post-war population growth was characterised by a different approach to urbanisation, it did not result in a dispersion of the population in low-density settlements. On the contrary, planning requirements reproduced an agriculturally oriented, open, relatively empty countryside and urban concentration.

The historical preservation of relatively abundant rural land dominated by agriculture provides much of the background to present discourses about rurality in the Netherlands. Much recent debate is about the future of this open, relatively empty space. On the one hand it is argued that future housing developments should be more dispersed in order to satisfy people's need for 'living space'. This option would eventually eliminate any rural-urban differences. Others argue that housing should be concentrated in high-density urban settlements in order to preserve the characteristics of the countryside and to avoid intensive long-distance travelling between home and work. Recent government policy documents seem to be in favour of concentration, thus supporting the model of a clear-cut morphological and functional separation between town and country. This is very much a continuation of past policies. However, rural areas are no longer predominantly defined as a territory for agricultural production. In the past, the transformation of rural areas was mainly structured by agricultural policy, which aimed at maximising production efficiency. More recently rural areas have become defined as cultural and natural assets, deriving their meaning from their capacity to satisfy urban demands for 'consumption space' in the broadest sense.

Over the past decade we have witnessed a sort of 'rural renaissance', a re-orientation of the significance of the rural in Dutch society. This transformation in thinking about rural areas took shape in the 1980s under the influence of growing urbanisation and the modernisation of agriculture. Agricultural modernisation was increasingly associated with environmental pollution, the degradation of landscapes and excessive budgets for farm subsidies. At the same time urban, industrial, and infrastructural developments increasingly occupied formerly rural areas. The countryside was no longer taken for granted as the unrestricted domain for agriculture, and the impact of urban development raised people's consciousness of rural areas as a limited good. Overall, social demands for access to rural areas for recreational and residential purposes developed simultaneously with more respect for the character of rural landscapes, nature and wildlife. The combined effect of these trends has been a gradual re-orientation of policy and planning and a redefinition of the countryside, culminating in recent policy documents that explicitly announce the diminishing significance of agriculture for rural areas. Rural policy discourses now focus on a multifunctional development of the countryside where nature, landscape, agriculture, recreation and other rural activities are in harmonious balance, serving the needs of both urban and rural residents. Eliminating the antagonism between development and preservation, between economic vitality and heritage conservation is seen as the main challenge.

The changing interpretation of rural areas in government policy closely corresponds with broader European policy developments. In the CAP 2000 working document *Rural Developments,* the DG for Agriculture of the European Commission (1997) describes rural areas as 'home to a great wealth of natural resources, habitats and cultural traditions' (p. 6). Rural areas, according to this document, 'have the potential to improve the *quality of life* for the whole of society, by providing an environment which is healthy and secure, with a high level of social integration and safety'. It goes on to say that many rural areas boost landscape amenities, the proper development of which can provide a key to local economic prosperity. The same applies to the wide potential offered by rural areas for recreation and the pursuit of leisure activities. Furthermore, rural areas provide a context for the production of quality products, with a well-defined identity and a traditional, cultural value.

New ideas about planning, regulation and governance accompany the discovery of the potential richness of the countryside, both for rural people and for society at large. Again, developments in the Netherlands seem to be in line with European policy guidelines. Policy documents recognise that the countryside across Europe and within national boundaries shows a great variety. This has important consequences for the kind of rural policies necessary to promote the vitality and quality of rural areas. National and European policies thus increasingly limit themselves to the formulation of broad criteria, which offer guidelines for regional and local initiatives in close collaboration with the people involved, a shift that can be clearly detected from the most recent policy document issued by the Dutch Ministry of Agriculture, which states that rural development is not a policy field that should be developed by central authorities, but should be based on locally developed plans that are initiated and implemented by the people themselves. This bottom-up perspective corresponds with Council Regulation (EC) No 1257/1999 of 17 May 1999 on support for rural development from the European Agricultural Guidance and Guarantee Fund (EAGGF) in stressing the geographical level of approach. According to the Cork Declaration, policies must have a clear territorial dimension and provide the 'framework for self-sustaining private and community-based initiatives', and be based on 'participation and a "bottom-up" approach, which harnesses the creativity and solidarity of rural communities' (p. 73). In summary, rural development must be local and community-driven within a coherent European framework. Individual solutions must be found for each region in the light of its inherent characteristics.

The Dutch planning system, however, is embedded in a politico-administrative structure that is essentially based on central planning with limited regional and local political power. Since the Second World War state formation has resulted in a gradual decline of political autonomy at the local level. This implies that the present shift in thinking about the significance of rural areas cannot easily be translated into different approaches to governance and political participation. The logic of this planning system has also engendered a rather mono-directed kind of rural discourse. The everyday experience of rurality and urbanity are not at the centre of the debate but instead scientifically constructed problems, which can easily be translated into long-term planning objectives. This chapter will argue that the present transformation in thinking about the future of rural areas has not yet found its counterpart in new political and administrative structures, and that rural discourses are still far removed from including rurality in its broadest sense. The character of this chapter is mainly descriptive. It describes how the countryside developed from a policy domain dominated by agricultural interests to an integrated rural policy domain and how spatial planning gradually incorporated the countryside into the wider context of urban development. A short sketch of the Dutch political and administrative system precedes our discussion of spatial and rural planning. We will try address questions concerning the extent to which local and regional authorities can play a role in a policy field that is strongly dominated by central government. This growing attention for rural development problems is accompanied by an intense debate about the future of rural areas. We argue that these discourses do not represent a break with the past, but continue some important themes.

We also argue that Dutch discourses on rural areas do not show many signs of considering the social and cultural characteristics of rural areas, even though new forms of governance and policy formation are not neglected. We conclude our chapter with a description of political changes at the local level, which allows us to answer some questions concerning the potential role of decentralised forms of policy-making in the Netherlands.

Administrative Structures

The Dutch political and administrative system is based on three levels of government: national, provincial and local (municipalities). The national government consists of the prime minister and his ministers, each representing a policy field institutionalised in national ministries. The composition of the government depends on the political power relations in parliament, which is formed every four years after general national elections. As none of the political parties has the majority, coalition governments are always formed after a long period of negotiation. Although the Netherlands has a decentralised political system, the central state decides about national policies in all fields, with different degrees of regional and local autonomy. At the regional level the country is divided into twelve provinces, each with a provincial government and parliament, which are composed on the same principles as the national government (with elected members). As the Netherlands is not a federal state, these provincial governments are in fact complementary to the national government in the sense that they are responsible for implementing national policies. Apart from such policy fields as taxation, foreign policy and defence, this implementation allows for some degree of autonomy and freedom to define additional measures. At the local level the administration is in the hands of municipalities. These municipalities are governed by a mayor (appointed by central government) and a board of aldermen and women, reflecting the composition of the elected municipal council.

The municipal level has been subject to drastic reorganisation over the past decades. Small municipalities have merged together, resulting in a decreasing number of much larger administrative units. Between 1980 and 1998 the number of municipalities went down from 811 to 548. Municipalities with fewer than 5,000 inhabitants were particularly affected by this trend (from 246 in 1980 to 24 in 1998). This reorganisation of local government has important implications for the level of the local population's involvement with policy. It has enlarged the distance between citizens and politicians, and resulted in an overall professionalisation of local civil servants. One reason for municipal rearrangement was to create more effective forms of administration, which was considered necessary in order to cope with the growing complexity of local government.

According to Herweijer (1999), mutual relations between government levels in the Netherlands are based on the powerful role of the state. Thus, provincial and municipal policies can in principle be overruled and the implementation of national policies be imposed. Furthermore, in order to implement national policies the lower administrations are financially very dependent on national resources. While the national government already controls many decisions made by the lower authorities, local and regional policy discretion is increasingly curtailed by legal rules. However, to understand Dutch policy-making, especially in the domain of spatial planning, it is important to realise that final decisions are taken at the central level and are thus binding, but that the preparation of policy decisions involves a broad national debate, involving all layers of government and interest groups.

Rural Development Policy and Planning

In the Netherlands, rural development policy is an integrated part of a larger policy domain involving the spatial planning of all land-use related activities and sectors, such as agriculture, housing, industry, infrastructure and so on. In a densely populated country such as the Netherlands, space is a scarce resource with multiple and sometimes opposing claims. It is therefore not surprising that the state centrally coordinates the spatial dimensions of policy decisions.

National spatial policy is based on three main political documents: the National Spatial Planning Document, National Sector Structure Plans (e.g. for traffic and transport, rural areas etc.) and the National Spatial Planning Key Decision. The first two instruments set out the main medium- and long-term policy principles and guidelines. A sector structure plan is primarily an instrument for sector planning (and as such drawn up by the appropriate ministry), but its content is also intended to facilitate an optimal coordination between sector planning and spatial planning. On the basis of these documents the government formulates its national spatial planning key decisions, which are subjected to and finally approved by both chambers of the parliament. Of course, this democratic legitimisation gives the Key Decisions important political prestige, but until the recent past they were only indicative, not legally binding. From January 1994 however, key decisions on certain projects of national importance could be made binding, (such as decisions on major roads, national railway links and major waterways). This policy framework enables the national government to take important decisions (e.g. subsidising major public works) and to directly or indirectly influence other public bodies (such as provinces, municipalities and water boards) to act in accordance with national policy. If the national government wants to change or influence provincial or municipal spatial planning policies, it can do so under the auspices of the Spatial Planning Act. According to this Act the national government has the function of supervising lower-level authorities. This supervising function has a wide scope, because it can be based on general policy goals and is not necessarily based on the parliamentary approved statements in the Key Decisions. This shows that national power is rather strong, notwithstanding the decentralised national political system.

This does not mean, however, that local authorities are powerless. Political party connections and pressure and interest groups put the national government permanently in a position of mutual interdependence with lower governmental bodies, which have a democratic legitimisation function as well. That is why the system of administrative consultation is very well developed during the process of policy formation. For that reason spatial planning procedures generally take a long time. In fact the period between the first publication of a national spatial planning document and the final approval in parliament may easily take five years, sometimes even more. Such a long policy-making process is caused by procedures such as extensive publicity, public participation (when anyone can respond), recommendations by official advisory boards, administrative consultations and finally the public discussions and judgements in the parliament. It is often the case that after the key decision is approved, the debate has progressed to such an extent that preparations for a new proposal are well underway. The open character of spatial policy preparation and the involvement of all political parties, administrative levels, interest groups and professionals, implies that there is a constant flow and exchange of ideas and plans, which makes it sometimes difficult to separate the virtual reality of discourse from concrete developments in rural areas.

Until the end of the 1970s, spatial planning was mainly concerned with housing and mobility (infrastructure) issues. However, in the Third Spatial Planning Document a special volume was devoted to the spatial dimension of rural areas (1979). The national government used a zoning principle as its main instrument to allocate functions to rural areas. The non-urban realm was divided in four categories: zones for agriculture (A); zones for agriculture, nature and other functions in large spatial entities (B); zones for agriculture, nature and other functions in small spatial entities (C); zones for nature (D). Furthermore, it was indicated which parts of the country were legitimated as zones with open landscapes. Where these areas were under urban pressure, a policy was formulated based on restrictions and constraints for growth and the spread of population. This policy of restrictive planning for open areas was continued, broadly speaking, in the Fourth Report. Partly because the restrictive policy is not popular with the public in general or among local administrators, it distinguishes between national severe restrictive areas and provincial rather restrictive areas (Mastop et al., 1996). In the national areas absolute limits to development have to be established around existing centres. The best-known restrictive area is the Green Heart of Holland, which is subject to the most severe planning restrictions. The area is defined in detail and no new developments are allowed within these limits. In provincial restrictive areas, the expansion of urban functions is permitted only within regional overflow areas. Even here, the province has to define ultimate limits to the built area for all rural settlements. Another instrument to keep the rural spaces as open as possible is the buffer zone policy. The goal of this policy is to maintain the open zones in between the urban regions through active central government investments in their development (supporting ecological and recreational use) and in some cases by purchasing land.

An important phenomenon in this Rural Areas Spatial Planning Document was an approach oriented towards regional issues. Around 1990 this approach was elaborated in the coordination of spatial planning and environmental policy in specific areas (Mastop et al., 1995). Such areas were distinguished in both urban and in rural regions. In rural areas, however, this regionally oriented policy category competed with several regional policy categories within the Ministry of Agriculture, Nature and Fisheries (national parks, national landscape parks, wetlands, valuable man-made landscape projects).

Rural spatial policy was developed in order to meet three challenges: to enhance spatial diversity, to protect and improve environmental quality, and to maintain the quality of life in rural areas. Three different sets of policy instruments attempt to achieve these objectives: the rural strategies, the agriculture and environment policy, and the quality-of-life policy. The rural strategies aim to offer prospects for spatial and economic development in the long term, and to create the conditions for sustainable development of the countryside through the harmonisation of spatial planning, water management and environmental policies. The rural strategies have to be implemented by other levels of government. For this reason, rural policy has to be incorporated in regional plans and takes its final legal shape in local land use plans. The second objective, to protect and improve environmental quality, has been approached by the instrument 'ROM-area-based policy', mentioned above. The quality-of-life policy includes provision to ensure that people do not become socially isolated just because they are physically isolated (Galle and Modderman, 1997, p. 28). Two ministries started a coordinated four-year quality-of-life project in eight selected rural regions (out of a total of six in the north, one in the southwest and two in the east of the country). The results were mainly psychological. All residents became 'aware' that their region was 'alive', although employment, shops, service, public transport still declined. Tangible effects included initiatives for recreational projects and for new types of services and cultural activities. In 1995 the two ministries stopped promoting these projects, because the regions involved could apply for subsidies from the European Regional Development Fund.

Until the 1970s, rural regions were primarily considered as agricultural production areas. Rural welfare was simply equated with increasing labour productivity in agriculture. In 1975, however, a programme was introduced for turning agricultural lands into nature reserves or 'nature management areas' providing financial compensation for applying environmentally beneficial farming methods. In the same period, issues of agricultural productivity became disputed because of growing surpluses (of meat, milk and cereals) and because of increasing awareness of agriculture's polluting effects. Competing claims from nature conservation, recreation and the protection of water quality gained power. From the mid-1980s onwards, policies for rural regions, stemming either from the two core ministries (Agriculture, and the Environment) or from provincial and local authorities, were increasingly concerned with interests other than agriculture (Frouws and van Tatenhove, 1999).

The two most recent Cabinets have set this priority also in political key-statements titled 'Dynamism and Innovation' (1994) and 'Power and Quality' (1999). In these political documents of the Ministry of Agriculture, Nature Management and Fisheries (ANF) the national government is showing responsiveness to many new trends in the field, but reacts with an orderly retreat of governmental responsibility. The cabinet sets the main policy reference points, facilitates new procedures and promotes a lot of innovative projects. However, direct responsibility for innovations is transferred from the national government to regional and local authorities and particularly to social actors and individual entrepreneurs. Nevertheless the Ministry perceives itself as the main coordinator from 'Brussels to Borculo' i.e. from the global to the local. Furthermore, the Ministry of ANF also has the task of coordination at the horizontal level, i.e. between the different Ministries at the national level concerning all affairs in rural renewal. In terms of its frames of reference, the cabinet promotes a regionally oriented approach. In cooperation with the provincial authorities and perhaps some other participants, the national government is in favour of the foundation of Regional Centres for Rural Renewal. These centres should develop regional knowledge and information networks. The provinces have to prepare regional strategies, which will be implemented on the basis of territorially oriented covenants between the national and the provincial administrative level (this is a new administrative instrument in this field).

Dutch rural policy has not only been redirected by national developments. The international and European contexts are equally important. The further liberalisation of trade in agricultural products (resulting from WTO agreements and the reforms announced by the EU Agenda 2000 proposals) strengthen the idea that the future of rural areas cannot be based on a further expansion of highly subsidised, export-oriented agriculture. The new EU emphasis on the quality of rural areas, which is expressed by a redirection of subsidies towards environmental management and rural renewal, corresponds with Dutch initiatives to develop an integrated policy on rural areas. Because of the specific characteristics of Dutch rural areas, the implementation of Agenda 2000 and the European Agricultural Model has not been received with much enthusiasm in the Netherlands. According to EU definitions, the Netherlands have hardly any rural areas and if EU regulations and subsidies only apply to large-scale open rural areas, the Netherlands cannot expect much. In recent advice to the government (RLG, 2000) it has been argued that the Netherlands is much better off with a more fine-tuned definition of rural areas, taking into account those areas that are under heavy urban pressure and are especially valuable for providing a contrast with urban life. The EU should extend its attention to ecologically poorer, densely populated areas, especially in those areas where quality improvements in space, nature and scenery is most urgent.

Dutch rural policy and planning now involves a multitude of centrally coordinated national and region-specific plans. At the national level the most eye-catching development concerns the creation of an ecological zone, which crosscuts the country and is intended to connect a multitude of environmentally sensitive areas. At the regional level a large number of so-called valuable landscapes have been designated as areas where special attention is given to the preservation of natural and cultural heritage. Recently many Dutch towns have taken the initiative to create more intensive links with the surrounding countryside in order to organise town-country relations at the micro level. The city of Rotterdam, for example, has developed intense communication networks with the rural area at its southern border, while such cities as Deventer and Leeuwarden increasingly incorporate neighbouring municipalities in city development planning. These initiatives undoubtedly mark the beginning of the definitive emancipation of the countryside from agricultural policy, and the start of a process of an integral rural-urban planning perspective, involving both urban and rural people.

Rural Discourses

Views on the countryside crosscut familiar political divides in several ways. According to Frouws (1998) rural discourses in the Netherlands can be divided into three main categories. First he distinguishes the *agri-ruralist* discourse. This discourse emphasises the central role of farmers in the countryside renewal process. According to this discourse, the value of the countryside, such as the landscape, open space, natural resources and heritage, are viewed as a result of the co-production and cooperation between man and nature. Farmers therefore have an essential role in preserving this identity and in meeting the social demands for an attractive countryside and high-quality food products. Self-regulation and endogenous development are of essential importance in the revival of rural areas and the establishment of a new identity between local actors and their environment. This discourse about rurality is a modern version of a discourse about the idealistic, self-sufficient nature of rural communities; this modern version has now opened up to serve the needs of urban society.

In emphasising the social dimension, this way of thinking differs from the *utilitarian discourse,* which focuses entirely on the economic dimension. The protection of the countryside obstructs opportunities to take advantage of its economic potential. By completely integrating rural space in the dynamics of modern markets for, for example, housing, recreation, quality food, high-tech agriculture and attractive landscapes, it can better meet growing demands for living, working and recreational space. According to this neo-liberal discourse, the commoditisation of rural areas and the breaking down of artificial boundaries between city and country are the best guarantee for meeting the diversity of social needs for quality of life. According to Frouws, this discourse represents a long-standing Dutch tradition of reclamation, exploitation and commercialisation of rural land.

The third discourse mentioned by Frouws is the *hedonist discourse.* Here, the emphasis is on the cultural dimensions of the countryside, in particular its scenic beauty and natural values. As such it plays an important role as a counter-image of congested urban life. This discourse is rooted in nature conservation movements and elite notions of the countryside as the garden of the city. The role of the state and derived forms of local governance are mentioned as the most important regulative mechanisms.

Frouws admits that his classification does not reflect any real divisions between interest groups or parties; it is rather an analytic construction of coherent sets of thought based on an analysis of texts such as government documents, in which consistency is sought after rather than the presentation of pragmatic choices. Frouws's division of discourses may in fact be read more fruitfully as a collection of themes that may coexist in different combinations. For the purpose of this chapter it is pertinent to discover the basic underlying themes and assumptions in official rural discourses in general. In doing this, we have to keep in mind that the importance and character of the rural may be expressed in a variety of ways. At one end of the spectrum the rural is defined by objective quantitative criteria and administrative units are classified accordingly into an urban–rural scheme (for instance the OECD classification); at the other end people develop images of the rural in everyday popular discourse. Statistical operationalisation and people's everyday perceptions of the rural only differ to the extent that they are located in different practical contexts – a scientific, policy-oriented context, and an unreflective popular one. Both are forms of representing rural reality, and both have implications for political or everyday practice. Within rural communities, policy fields, arts and literature there seems to be a plurality of ruralities, associated with a variety of lifestyles, ideologies and interests.

According to Hoggart et al. (1995), however, the character of rural discourses unmistakably bears the imprint of a specific national context. While the rural as a specific cultural theme takes a crucial role in Scandinavian countries and the British Isles, it is almost absent in Mediterranean countries. The national context provides a historical cultural and political background against which concepts of rurality appear as mere variations on a basic theme. Thus in English research the 'rural idyll' takes an important place. In France, concepts such as *terroir* and *paysan*, and in German areas *Bauerntum* and *Heimat* are central concepts. Apart from Schama's *Landscape and Memory* (1995), a historical study about the meaning and representation of landscapes, there are no comparative European studies that link national rural discourses to the historical and cultural embeddedness of the rural. In the Netherlands there are only partial studies, which trace the representation of nature (Windt, 1995) or rural culture (Schuurman, 1991; Van Ginkel, 1998), albeit from quite different perspectives.

Rural discourses in the Netherlands, as far as they appear in national public debates, are basically about a new relationship between the rural and the urban. The construction of these new relationships does not take place in a historical and cultural vacuum. New paradigms about rural–urban relations are thus partly based on long-standing ideas about planning and intervention, and obviously cannot escape from deeply embedded cultural ideas about rurality. In a sense, present ideas about the restructuring of rural areas represent a mixture of the classical modernist and postmodern conceptions of rural values. The modernist legacy is explicit in the firm belief in the planning and intervention capacities of the state, while the postmodernist element is exemplified by the reinvention of the rural.

The creation of landscapes is a fundamental aspect of Dutch history. Most agricultural land was reclaimed from a semi-natural state in several phases in history and successively adapted and improved. This reclamation process was not simply done by peasants who gradually cleared forests and moors, but was mostly executed by commercial companies and institutions, and involved huge inputs in capital and organisational skills. The belief in the capacity to control nature, and to design space according to specific functional, commercial, and political requirements is still omnipresent today. Although rural planning is no longer dominated by the functional requirements of agriculture, the belief that rural areas *can* be designed according to specific needs has not disappeared. The reorientation of planning policies simply involves a redefinition of 'rural products' and the development of guiding images oriented towards the future, while maintaining the idea that large-scale interventions can create a new balance between functional requirements and socioeconomic needs. Rural planning and policy practices may have shifted considerably from production to consumption-led principles, but there has been no significant change in the underlying belief of landscape malleability and the idea of functional requirements.

Creating a new rurality, however, involves a break with the definition of the countryside as a productive area and a widening of its function from the provider of food to the provider of social and cultural products. This redefinition of the countryside differs from the post-war modernisation project, when structural policies were aimed at eliminating rurality as a social and cultural element in Dutch society. The praise of the virtues of numerous small farmers and their supposed contribution to the cultural richness of society was rejected as a dangerous ideology, hampering the war against poverty and backwardness (de Haan, 1993). Basically, the concept of rurality (*platteland*) was gradually stripped of all its social, cultural and historical connotations, becoming a purely functional concept related to agricultural land use. The modernisation project was indeed designed to set in motion a civilising process. Tradition, community ties and belonging, regional differentiation and so on were all seen as obstacles to the creation of a national welfare state. Participation in welfare and education was conditional on the cutting off of all ties with localism, regionalism and historical patterns. This was given its greatest expression in the fields of agricultural policy and land use planning on the one hand and regional policy on the other. Evert Willem Hofstee, the first professor of rural sociology in the Netherlands, introduced the theoretical concepts for this modernist project (de Haan and Nooij, 1985; Nooij, 1970). He basically argued that modernisation was based on the rejection of the past (tradition) and the acceptance of a cultural pattern based on rational decision-making. For agriculture this implied the eradication of farming as a way of life and its replacement by modern entrepreneurs. The landscape needed to be tailored in accordance with efficiency and productivity criteria, and the inhabitants of rural communities had to give up their attachment to a specific place and orient themselves instead towards national culture. The cultural and social 'urbanisation of the countryside' and the modernisation of agriculture would together result in the end of rurality and with it the end of rural sociology. Agriculture would become a sector like all others, and rural communities would be integrated into urban society.

By the 1980s this concept of rurality met with increasing criticism. In contrast to the pre-war period, when concern about the decline of the countryside was predominantly related to the social and cultural disintegration of rural life, the emphasis was now mainly on 'spatial quality'. In the pre-war period, rural civilisation was seen as the source of vitality for society at large, while from the 1980s onwards, the quality of rural areas had to be improved and protected to meet urban demands. The classical modern period between the 1950s and 1980s can be seen as an attempt to isolate the rural spatially and functionally from urban economic activities, and to deny the potential multifunctional use of rural space. During this period the social and cultural components of living in rural areas had successfully been abolished from the policy and research agenda.

Dutch rural policies aim at creating rural scenes, which are clearly differentiated from urban places, but which are completely integrated into urban society. Although the post-war modernisation period was characterised by the denial of the cultural meaning of the rural and by maintaining a spatial separation from the urban economy, there are also similarities with present policy attitudes. One concerns the already mentioned belief in planning and intervention; the other concerns the lack of interest in the social and cultural rural landscape. In order to illustrate this last point we will contrast two types of rural problems.

The Experience and Construction of Rural Problems

'Charivari in Woudenberg after protests.' 'Incomers want to get rid of wild parties.' These two headings announce the story of a conflict in a rural community in the centre of the Netherlands (Determeijer, 1999). Some 10 years ago, a local farmed started to organise parties for rural youth in a shed originally built to house pigs. At the time legal measures forced him to reduce hog capacity and he began to look for alternative uses for his buildings. The concrete building, in a sparsely populated rural area, was an excellent location to organise 'shed parties' (*schuurfeesten*), a popular form of entertainment for rural youth. There, in an orgy of beer, dance and music, rural youth could let off steam without restriction and without being a nuisance to other villagers. But for the immediate neighbours, two families with an urban background enjoying the quietness of country life, these parties were an enormous annoyance. They claimed that such activities did not belong in a rural area. After continuous complaining without conclusion they decided to take legal action, which resulted in a formal prohibition of the parties. Given that it was known that the neighbours had taken legal steps, they were confronted with violent attacks on their property and constant intimidation, to such an extent that one family had to take refuge. According to the mayor, who intervened for a peaceful settlement, taking legal steps was a big mistake: 'They should have had a nice drink together instead.' He thought that the opposition between town and country had played an important role in the conflict, referring to incomers who did not understand rural communities and thought that they could impose their own principles.

This case illustrates in a nutshell the character of 'rural problems' in the Netherlands. Limitations on intensive pig farming motivated an innovative farmer to use his farm capital in an alternative way. Rural policy measures stimulate this sort of behaviour in all kinds of ways (see Van Broekhuizen et al., 1997). His activity met with a favourable response amongst the rural youth, who needed a place of their own where they could express their identity. Some studies (see de Haan, 1998) have shown that rural youth in the Netherlands have been at the forefront of opposing urban influences in rural areas. An emerging rural youth subculture thrives on peasant symbols such as overt lumpen physical and verbal behaviour, local idiosyncrasies and xenophobia. Rural identity is one of the slogans in policy documents, and although it is unclear what is understood by it, it is certainly not meant to refer to forms of localism that reject openness to external influences. In a recent policy document by the Ministry of ANF, *Voedsel en Groen* (2000), rural–urban social and cultural differences are denied, but at the same time it argues for a strengthening of rural identity. The underlying urban-rural differences in the conflict clearly demonstrate the diversity of rural representations at the local level. Incomers seem to have certain expectations about living in the countryside which contrast with the practices of other local residents. By taking legal action the incomers demonstrated their position as outsiders, unable to settle their disputes in an informal way.

The image of rurality described in policy documents seems to be far removed from everyday reality in rural communities. Those documents represent rurality in terms of morphological and functional criteria, hypothesising the demise of a social and cultural divide between country and city. New claims on the countryside are not pictured as an urban infringement on rural space, but as evolving from general social, economic and cultural needs. There may be differences of opinion about the spatial ordering of agriculture, tourism, residence, nature and so on, but these differences are not seen as based on cultural contrasts between urban and rural lifestyles. However, not all rural conflicts are embedded in the structure of rural community life, as is shown by the next newspaper report.

'New housing estate threatens meadow birds' is the heading of an account about a conflict in a Friesian village (de Mik, 1999). The contested object is the Bullepolder, an area of about 120 hectares of grassland not far away from the provincial capital of Leeuwarden. The inhabitants of the village are furious, because the town of Leeuwarden under which the village is administered, has bought the land from two farmers for 10 million guilders. The town needs this land in order to distribute it to potential house-owners who want to combine living in the countryside with being near the city. The allotments are designated particularly for high-income groups. The town council's motivations for acquiring land for housing is that it wants to stop the exodus of well-to-do inhabitants for whom there are not sufficient housing facilities in the town. There is a tendency for these people disperse over the rural areas in the province, which is considered as a threat to the vitality of the provincial capital. Several hundred villas are proposed in a new green suburban environment. The inhabitants of the adjoining village are fiercely opposed to the plan, however. Curiously enough they do not express their resistance in some sort of rural-urban discourse about locality and identity; the social and cultural dimension is in fact completely absent. Opposition is based entirely on a desire to protect birds. Building up the area would greatly damage the habitat of various bird species that build their nests in the fields. The villagers have even written a letter to the European Commission demanding the allocation of special status to the area. The opponents of the plan are extremely well informed about the behaviour of birds, their preferred habitat, their nesting behaviour and so on. However, sympathy for animals has insufficient legitimacy in the opposition to these plans. In order to be taken seriously, scientific data and international concerns have to be demonstrated. The opponents of the plan represent the interest of birds, and at a higher level an environmental movement that cannot be located in a specific locality or along rural–urban lines.

This local conflict is quite different from the one mentioned before. Most significantly, the opposing parties are not involved in personal conflicts. Rather, the conflict is an institutional one. Furthermore, the problems are not rooted in everyday local experience in a rural community, but evolve from plans projected onto spatial objects. In that sense, the problem is a political and environmental construction. The arguments used in the dispute are based on scientific data about future needs for housing and the ecological niche of rare birds. The conflict will no doubt be settled during a civilised meeting where policy makers promise to re-evaluate the situation on the basis of materials provided by an independent research committee.

These two examples represent the scope of rural problems in the Netherlands in a nutshell. Although there are great differences, both concern the ordering of rural activities and spatial arrangements in specific locales. As such they go to the heart of the basic questions facing rural areas in the Netherlands. However, the two are connected to quite different research traditions, which are not equally represented in the Netherlands: research focusing on social and cultural life in rural communities, and research centred on planning and policy-making. Although planning and policy-making research may also choose rural communities as a research site, the emphasis is always on an assessment of future-oriented, government imposed interventions. As such it may be very interesting, but it is logically limited by its legalistic and governance bias, which means that rural practices and interactions in the broader sense are neglected. There are also paradigmatic differences in the sense that rural community research is especially interested in locally embedded discourses and popular representations of rurality and urbanity, while policy and planning oriented research is most interested in capturing institutional discourses and practices.

Rural Restructuring, Leadership and Local Power

Rural areas only became a substantial issue in national spatial policy-making in the late 1970s. Until then, the rural realm was the power domain of the Ministry of Agriculture. For national spatial planning agencies, rural areas were out of their formal capacity and as such represented 'remnant space'. The creation of a specific Ministry of Agriculture (1945) was symbolic for the relevance and power of the agricultural policy community in the post war period. The Ministry of Agriculture was the main partner in a neo-corporatist regime that controlled public and private agricultural affairs. The harmony between the state and the agricultural sector created equilibrium in the power relations. In this balance of power (see Goverde and Hinssen, 1994; Goverde, 1995), the government received cooperation, information, a disciplined peasantry and legitimisation of its authority, and offered in exchange influence to the organised farmers interest groups, information, social standing for the farmers representatives and a monopoly in the articulation of their interests. The main 'rules' for the articulation and aggregation of interests were exclusiveness, consensus, depolarisation, promoting a technological approach and elitism (Frouws, 1993).

The leaders of the farmers' organisations often had their own farms and were recruited from the local branches of those organisations. That is why in the pre-war decades, as well as from 1945 till about 1975, leadership at the regional and for some representatives at the national level, was strongly interconnected with local power in rural areas. This regime of neo-corporatism was the organisational engine for the process of modernisation in rural areas and in agribusiness. The local farmers did not always understand the necessity of modernisation. Sometimes there was public resistance particularly when the farmers had to vote for a proposal of reallotment. In such cases it was often the task of local leaders, as representatives of the higher powers, to legitimise the implementation of regional and national policies and to discipline resisting local farmers. In such cases the interconnectedness between the different local pillars of power – local government, local political parties, the church and local farmers' leaders – was often expressed publicly. Briefly, the 'golden years' of the Green Front were characterised by an effective power structure, linking power holders at the central government level directly with farmers at the local level. Although indeed this structure was not in the interests of all farmers, those with an interest in rural development at the local level were actively involved in the political structure.

The dominance over rural areas by the Ministry of Agriculture and the persistence of a political regime in which farmers' organisations with strong local representation could successfully defend their interests partly explains the narrow vision of rural politics until the late 1970s. While planning agencies were mainly focusing on questions of urbanisation, infrastructure and mobility, rural areas remained outside the scope of planning objectives, which in a rural context were mainly defined from an agricultural perspective. The late appearance of the rural in the wider sense on the political agenda may also be related to developments at the local level. Although the farming population was in decline, and rural communities increasingly had a diverse population, these changes were not apparent in the composition of representative political boards in rural communities until the end of the 1960s. According to Munters (1989) farmers were significantly over-represented in local municipal councils in the first two decades after the Second World War. For instance, in 1958 the political position of farmers had not changed much since 1939. According to Munters, the 'ungreening' of local political power positions was 20 to 30 years behind the 'ungreening' of the rural community population. How did farmers manage to reproduce positions of power while rural communities were changing rapidly? Munters attributes this 'retarded revolution' to the fact that farmers continued to play an important role in local political parties. This role was not contested by newcomers and not affected by the shifting composition of the local population. As Verrips (1978) has shown in his study of a village in the west of the Netherlands, local political power was mainly based on traditional power resources such as land and extensive kinship networks. Each rural community had an elite group of local farmers who not only dominated local politics, but also enjoyed high moral prestige as leaders of the Church, the school, local organisations and water management boards. They were the 'natural' local leaders at a time when local communities had a high degree of autonomy and the bureaucratisation and professionalisation of local government were still weak. It was only with the massive inflow of incomers (see Brunt, 1974) and the reorganisation of the local administrative system that these local leaders gradually lost their moral power and were replaced by political leaders recruited from other social groups.

The breakdown of the 'Green Front' and the loss of power of the agricultural policy community were clearly related to the reconsideration of agriculture as the dominant factor in rural areas. With the discovery of rural areas as urban consumption space, new political power networks have been created, linking a wide variety of interest groups, civil servants, academics and politicians. Although this network does not represent a homogeneous bloc, its main characteristics are that a professional elite dominates it and that it has no clear linkages at the local level. Interest groups of public and non-political public organisations in such fields as nature and landscape protection are often organised at a national level and are thus not embedded in everyday rural practices, as was the case with farmers during the corporatist regime. Moreover, with the reorganisation of municipalities, local administrations have increasingly performed the role of representing the government instead of representing the local people by whom they were elected. The resulting (and growing) distance between local government officials and the population and the far-reaching professionalisation of civil servants has thus resulted in a minimal involvement of rural people in rural policies. Although they may participate in small projects, the main transformation in their rural environment is governed by rules, regulations and plans that have been initiated by extra-local policy decisions.

This bureaucratisation of rural development may be illustrated with reference to experiences with LEADER projects in the North of the country.[1] In 1990 the European Commission introduced the LEADER programme, which was intended to stimulate regional endogenous development and the participation of local groups and governments. LEADER I (1990–94) only applied to Objective 5b areas and the Netherlands hardly benefited from the new initiative. Under the LEADER II regime more regions could apply for funds, but the Ministry of Agriculture decided to concentrate the available finance in only four regions. In order to assess the extent to which the municipalities and inhabitants at the local level were involved and the effects of the LEADER initiative on local political relations, we will briefly discuss the experiences with LEADER in the region of northwestern Friesland, which was involved in both LEADER programmes.

The initiative for participation was taken by the Province of Friesland. LEADER I and II were seen as providing opportunities to increase funding for a regional development plan that was already in progress. The European Commission's requirements were all fulfilled and LEADER was thus successfully integrated into existing policy arrangements. Unlike LEADER projects in other countries (where several local groups are involved and take the initiative) all activities were centrally coordinated by one group, composed of selected people who were supposed to represent the local population. However, decisions in this group were prepared completely by Provincial civil servants, who were also responsible for the implementation of the projects. Most of the projects were already decided before the start of LEADER, and the population had very few opportunities to suggest new projects. In general, the aims, problems and projects were centrally defined by provincial authorities. At municipal level, the village councils only had a say in the approval of the amount of co-funding. Overall, LEADER was very much a project in the hands of provincial civil servants. They organised popular participation and established contacts between parties. The main conclusion from the LEADER experience in Northwest Friesland is that the local population and its governments were formally represented in the coordinating group, but in practice the LEADER idea did not have any impact. Coordination of the project was mainly in the hands of the province, and the population's representatives in the coordinating group had no obligation to consult the people they represented. Many of the projects were planned at the provincial level, and based on political preferences in the provincial governments.

During the last decade, however, municipal elections saw the revival of local political parties, particularly in growing rural communities. These parties have no representation in provincial or national governmental bodies and are purely based on local issues. Perhaps this indicates rural people's dissatisfaction with the regime of national rules, decisions and policies, and distrust in established political parties. This is only one indication of the fact that with the integration of municipalities in a national political system, rural communities have not lost their identity. Rather, it seems that there is tendency at the local level to be critical of the issues with which they are confronted in the name of national interests. Local governance tries to liberate itself from being a dependant executive body following party political lines and national directives. Local interests take priority. In both national and supra-national governmental entities there seems to be quite a responsive attitude to these new demands for local self-regulation and determination. Among policy makers and political scientists this is reflected in an intensification of debates on participation and interactive policy-making, new ways of implementing policy and a decentralisation of decision making and political management. Perhaps a network or self-governing approach is strong enough to create a fruitful arena for negotiation between politicians, civil servants, professionals and (organised) citizens at different administrative levels. At the present time, it is only in open planning processes or interactive policy-making that the constitutional demand for cooperative politics between governmental levels and institutions can be fulfilled. Particularly for the governance of rural areas, the intentions of the cabinet in its last Ministry of Agriculture policy report are in accordance with this approach. How should rural development be organised? Whatever the concrete form of the organisation it should be based on values of openness, participation, responsiveness and accountability. Indeed, the regional level is probably the best equipped for process management. At this level the demands of political-administrative convergence at the (inter)national level can be best matched with the demands of cultural divergence at the local level.

Conclusion

This chapter has argued that ideas about the future of rural areas in the Netherlands show continuity as well as a fragmentation of values. On the one hand, production and market criteria (efficiency, growth, profit), which dominated rural areas until the beginning of the 1990s, can now not only be observed among agricultural interest groups, but have extended to other spheres as well. Leisure areas, nature, cultural-historical landmarks and landscapes are increasingly perceived as commodities and services, which can be exploited for investment and profit. On the other hand, the concept of multifunctional land-use accommodates a plurality of values, each representing specific views on the nature and meaning of rurality in a densely populated country. There is thus a great diversity of users, interests, images and policy-arrangements concerning the future of rural areas. In order to explain the fragmentation and conflicts of values in rural areas, it is important to distinguish between problems experienced in everyday life, and the way in which problems are perceived and constructed in governmental reports and governance processes.

Although the Dutch political administrative system formally grants a great amount of policy freedom, if not autonomy, to the lower governmental tiers (the provincial and particularly the municipal level), our description of the functioning of the system shows that the national government only allocates a limited degree of flexibility in the field of rural restructuring and development. However, over the past decade the implementation of rural policies has changed. Until the early 1990s a neo-corporatist system controlled rural areas. This system was based on the dominance of farmers at the local level and their representation in the agricultural policy community at the highest political and administrative levels. It is quite surprising that this system could keep its strength for such a long time. In fact, this was an effect of the so-called 'retarded revolution' in the 'ungreening' of rural communities in the Netherlands. Even today the transformation in thinking about rural development has not yet been translated fully into new political and administrative structures. However, the national key reports (1994 and 1998) on rural areas foresaw a more important role for local and regional authorities.

What conclusions can we draw from our chapter for the future politics of rural development? Both the global European contexts will be of great relevance for rural development in the Netherlands. If the WTO regime develops towards wider and deeper liberalisation of agro-food world markets, and if the EU is to be enlarged with the entry of countries like Poland, which can produce agricultural products closer to world-market conditions, the total amount of subsidies to Dutch farmers will decline seriously. Many farmers will attempt to adapt to these conditions by permanent rationalisation (i.e. the continuation of the modernisation project). Of course, this approach needs very innovative, very high-tech oriented entrepreneurial farmers. In general they will possess large farms in the main agricultural areas or will be located on agribusiness sites. However, many firms will not survive this economic competition in the long run. It is also expected that CAP facilities for cross-compliance, (i.e. subsidies for multifunctional farming) will help some other farmers to survive economically as well. These farmers will produce mostly eco-agricultural products in combination with additional activities in fields such as nature protection, small camping sites or care-farming.

From a spatial planning perspective it is our hypothesis that nationwide organised interest groups campaigning on nature and environmental issues will continue to press public authorities to enlarge natural areas, particularly those near the most densely urbanised areas, under the slogan of 'freeing the landscape'. However, the aims of these organisations will no longer be reached by emphasising nature alone. On the contrary, mainly for reasons of public legitimation these organisations argue for nature and environment to be integrated into the economic system. At the margins of these areas, there will be enough services and facilities (walking, jogging, biking, skating routes; information centres, restaurants), which people will use during leisure hours. In this way the visitors' behaviour in these nature areas can be controlled to a certain extent. The result will be a park landscape, particularly within the urban ring (Randstad Holland and its extension to Amersfoort-Zwolle, Arnhem-Nijmegen, Eindhoven and Breda). Outside this urban ring more open agricultural spaces will be found, although even there private investors will be ready to produce the so-called 'good life feeling' by creating golf courses, camping sites, fun parks and biking and hiking routes. In these areas, facilities for weekends and short holidays will be dominant. Local restaurants, farmers and environmentalists will offer their eco-products labelled for commercial reasons as typical for the region's culture. The very wealthy people (particularly captains of industry, hotel chains, retirement funds and life assurance companies) will establish new estates in both areas. These new developments will underline the image of park landscapes in these rural spaces.

Concerning local power in rural areas, we believe – although much more research has to be done – that a shift in the balance of power can be indicated. Rural communities are mostly composed of concentrated settlements and dispersed farms, located near their fields. During the neo-corporatist period local power was in the hands of these farmers, who directly and indirectly regulated rural space. However, at present local power is generally in the hands of the people living in the built-up areas. These non-farmer village people are local merchants, well-educated suburban habitants, former farmers (generally retired people), civil servants, business people etc. This shift in the balance of power at the local level thus has a spatial component. Although local municipalities have the authority and obligation to make decisions about their rural areas, the production of local spatial plans has been severely limited by the national government over the past decade. In fact, local authorities now have more influence over the built-up parts within their territories (housing, employment, schools, neighbourhood services, sports fields, public order, safety etc.) than over their open, (former) agricultural spaces where national and provincial authorities and quasi governmental institutions are increasingly responsible for management and planning.

In the long term, the shift of local power from the rural outskirts to the rural centres and the relative loss of local political influence on rural space are not the only relevant factors. Not only in terms of real power, but also in cultural terms the gap between rural policies and people's lifeworlds is increasing. While the broad political spectrum in the Netherlands is grappling with the concept of a multicultural and multi-ethnic society, rural policy and planning continue to implement countryside images that basically emerge from elite and scientific culture. To what extent can ethnic and other cultural groups and individuals in Dutch society identify themselves with the multifunctional images of rurality promoted by the cultural and political elite nowadays? And are market-driven concepts of rurality perhaps closer to people's ideas about the use of space? The government's and its planning institutions' view that the Netherlands is 'full' is mainly based on the assumption that the urban population needs open space, and that they cannot therefore further reduce its quality and quantity. For people living in compact settlements the feeling of 'fullness' is however based on limited private space and overcrowded streets. Rural people, on the other hand, are confronted with limited or no possibilities at all to build the necessary houses for an increasing number of villagers. These contradictory processes thus leave both urban and rural people unsatisfied. Perhaps the Dutch government should not define the Netherlands as a densely populated country which needs to protect its rural areas, but as a sparsely populated town with plenty of space to create agreeable living conditions.

Note

1   The authors would like to thank Froukje Boonstra for her insights and comments concerning the implementation of the LEADER projects in the Netherlands.

References

Broekhuizen, R. van, Klep, L., Oostindie, H. and Ploeg, J.D. van der (1997), *Atlas van het Vernieuwend Platteland: Twee honderd Voorbeelden uit de Praktijk*, Misset, Doetinchem.

Brunt, L. (1974), *Stedeling op het Platteland: een Antropologisch Onderzoek naar de Verhoudingen Tussen Autochtonen en Nieuwkomers in Stroomkerken*, Boom, Meppel.

Determeijer, B. (1999), 'Buitenstaanders willen af van woeste feesten', *NRC Handelsblad*, 25 March.

European Commission (1997), *Agenda 2000*, COM (97) 2000, Office for the Official Publications of the European Community, Luxembourg.

European Commission, Directorate General for Agriculture (DG VI) (1997), *Rural Developments*, CAP 2000 Working Document, European Commission, Brussels.

Frouws, J. (1993), *Mest en Macht*, Wageningen University, Wageningen.

Frouws, J. (1998), 'The Contested Redefinition of the Countryside: An Analysis of Rural Discourses in the Netherlands', *Sociologia Ruralis*, vol. 38, pp. 54–68.

Frouws, J. and Tatenhove, J. van (1999), *Regional Development and the Innovation of Governance*, Paper presented to COST A12 Working Group 4 meeting, 25–27 March 1999, Vienna.

Galle, M. and Modderman, E. (1997), 'VINEX: National Spatial Planning Policy in the Netherlands during the 1990s', *Netherlands Journal of Housing and the Built Environment*, vol. 12, pp. 9–35.

Ginkel, R. van (1998), 'Illusies van het Eeuwig Onveranderlijke. Volkskunde en Cultuurpolitiek in Nederland, 1914–1945', *Volkskundig Bulletin*, pp. 345–84.

Goverde, H.J.M. (1995), *Macht om het Maaiveld. Een 'Koude Oorlog' Politicologisch Beschouwd (Power around the Surface)*, inaugural professorial lecture, Wageningen University.

Goverde, H.J.M. and Hinssen, J.P.P. (1994), 'Machtsbalansanalyse', in L.W.J.C. Huberts and J. Kleinnijenhuis (eds), *Methoden van Invloedsanalyse*, Amsterdam/Meppel, Boom, pp. 98–119.

Haan, H.J. de (1993), 'Images of Family Farming', *Sociologia Ruralis*, vol. 33, pp. 147–66.

Haan, H.J. de (1998), 'Jongeren Tussen Stad en Platteland: Droom, Werkelijkheid en Mythe', in Stichtin AIR (editorial collective), *Waar het Landschap Begint* (*New Landscape Frontiers*), Architecture International Rotterdam, Rotterdam, pp. 57–66.

Haan, H.J. de and Nooij, A.T.J. (1985), 'Rural Sociology in the Netherlands; Developments in the Seventies', *The Netherlands' Journal of Sociology*, vol. 21, pp. 51–62.

Herweijer, M. (1999), 'De Relaties Tussen Rijk, Provincies en Gemeenten', in H. Daalder, W. Derksen and A.P.N. Nauta (eds), *Compendium voor Politiek en Samenleving in Nederland* (May), Bohn Stafleu Van Lochem bv, Houten, Diegem, Deel C, pp. C0500–1/30.

Hoggart, K., Buller, H. and Black, B. (1995), *Rural Europe: Identity and Change*, Edward Arnold, London.

Mastop, J.M., Goverde, H.J.M., Verhage, R.W., and Zwanikken, T.H.C. (1996), *Ervaringen met Restrictief Beleid. Doorwerking van het Restrictief Beleid uit de Vinex op Provinciaal Nivea*, VROM 95604/b/1–96, RPD-publikatie, Den Haag.

Mastop, J.M., Leroy, P., Goverde, H.J.M. and Geest, H.J.M. van (1995), 'De ROM-gebieden Aanpak Funktioneel Ongeregeld', *Stedebouw en Volkshuisvesting*, vol. 76, pp. 9–14.

Mik, K. de (1999), 'Nieuwbouw Bedreigt Weidevogels', *NRC Handelsblad*, 24 March.

Ministry of Agriculture, Nature Management and Fishery (1994), *Dynamiek en Vernieuwing* (*Dynamics and Innovation*), The Hague.

Ministry of Agriculture, Nature Management, and Fishery (1999), *Nota Kracht en Kwaliteit* (*Power and Quality*), The Hague.

Ministry of Agriculture, Nature Management, and Fishery (2000), *Voedsel en Groen* (*Food and Green Spatial Areas*), The Hague.

Ministry of Housing and Spatial Planning, (1979), 'Derde Nota over de Ruimtelijke Ordening, Nota Landelijke Gebieden (Third Report on Spatial Organization in the Netherlands. Tome 3: Report concerning Rural Areas)', *Tweede Kamer*, no. 14 392, 1978–79, 22.8.1979, nos 1–2, 9.

Munters, Q.J. (1989), *De Stille Revolutie op het Agrarische Platteland: Boeren en Openbaar Bestuur 1917–1986*, Van Gorcum, Assen.

Nooij, A.T.J. (1970), 'Rural Sociology in the Netherlands', *Sociologia Neerlandica*, vol. VI, pp. 168–74.

RLG: Raad voor het Landelijk Gebied (Council for the Rural Areas) (2000), *European Integration and Regional Diversity*, Amersfoort.

Schama, S. (1995), *Landscape and Memory*, HarperCollins, London.

Schuurman, A.J. (1991), 'Tussen Stereotype en Levensstijl: De Ontwikkeling van de Plattelandscultuur in de Negentiende Eeuw', *Tijdschrift voor Geschiedenis*, vol. 104, pp. 532–47.

Verrips, J. (1978), *En Boven de Polder de Hemel; een Antropologische Studie van een Nederlands Dorp*, Wolters Noordhoff, Groningen.

Windt, H.J. van der (1995), *En Dan: Wat is Natuur Nog in dit Land?: Natuurbescherming in Nederland 1880–1990*, Boom, Amsterdam.

Chapter Four

Influences on Leadership and Local Power in Rural Britain

Rachel Woodward and Keith Halfacree

This chapter presents an analysis of the factors influencing leadership and local power in rural Britain.[1] Issues of leadership and local power are important because they shape economic and social development trajectories; public funds spent in rural development rely ultimately on implementation at the local level, and local support and involvement are increasingly recognised as a critical component of rural development. Whilst much rural sociology has traditionally mapped patterns of social stratification and power relations, and much rural geography been concerned with the analysis of economic development patterns and policy mechanisms, in more recent times and as a consequence of policy developments, attention has been turned to the analysis of local power as part and parcel of a wider investigation of both social and economic change in rural areas.

The chapter starts by providing some basic contextual information on rural Britain, considering socio-demographic changes and economic developments. Consolidating this background overview, the second section provides a brief overview of the principal administrative structures governing the management of rural areas at national, regional and local levels. The third section provides a description of recent European Union (EU) and national policy initiatives for rural economic and social development. The focus, in particular, is on those programmes which have sought to incorporate elements of popular or local participation within wider existing political and administrative frameworks. The fourth section highlights the discursive context in which rural development policy is made and suggests the role of these selective discourses in shaping new governance practices within rural Britain. The fifth section examines the roles of different groups of actors within the rural development process. This includes the roles of old and new elites, and 'insiders' and 'outsiders', and the influence of political discourses of regionalism and gender equality in shaping economic and social development trajectories in rural areas. The final section concludes by raising some basic questions and themes of rural development programmes and empowerment.

The Socio-demographic and Economic Structure of Rural Britain

The socio-demographic and economic structure of rural Britain provides an initial context for later discussions of local power, leadership and restructuring in rural communities. Any account of the dynamics and structure of rural Britain as it enters the next millennium must begin with the population changes which have restructured the countryside sociologically over the post-war period. Arguably, these changes have transformed the space of the British countryside from one dominated by issues of primary production to one where issues of consumption are seen to be especially significant. This transformation is captured in the idea, widely deployed by British academics, that we are moving towards a more 'post-productivist countryside' (for example, Marsden, 1998).

The Geography of Demographic Change

Evidence of significant demographic change can be seen quite simply in Table 4.1, which shows British population changes 1981–91 with respect to different categories of local government districts.[2] The table shows a strong negative correlation between the degree to which a district was urbanised and its population change. This corresponds to Fielding's (1982) definition of counterurbanisation, the key demographic trend within much of the developed world (see Champion, 1989; Rees et al., 1996). Overall, in spite of a much trumpeted revival of the population of London throughout the 1980s (Champion and Congdon, 1988), Table 4.1 shows that it is the three bottom categories in the table ('Resort, port and retirement districts', 'Urban and mixed urban/rural districts' and 'Remoter, mainly rural districts') – corresponding to the more rural areas – which have been gaining population most, in both absolute and relative terms. Thus, trends since the 1970s have been largely maintained.

Sticking with the counterurbanisation theme, it is a geography of net migration which is largely responsible for bringing about the contrasts shown in Table 4.1. This is demonstrated further in Table 4.2, where a slightly more rural-sensitive breakdown of districts shows the contribution of net migration to rural growth. Note here, in particular, how it is the 'Most remote rural' category, distant from the commuting influence of the large cities, which is gaining population most rapidly. Whilst Table 4.2 clearly shows that a lot of

Table 4.1   Population changes (millions), 1981–91, by district type

	Population 1981	Population 1991	Change (thousands)	Change (%)
Inner London	2.55	2.57	+16	+0.6
Outer London	4.26	4.24	-19	-0.4
Principal metropolitan cities	4.32	4.09	-235	-5.4
Other metropolitan districts	8.70	8.58	-122	-1.4
Large non-metropolitan cities	3.67	3.63	-44	-1.2
Small non-metropolitan cities	1.92	1.96	+35	+1.8
Industrial districts	7.44	7.56	+125	+1.7
New Town districts	2.55	2.74	+188	+7.4
Resort, port and retirement	3.37	3.63	+266	+7.9
Urban and mixed urban/rural	9.84	10.40	+561	+5.7
Remoter, mainly rural	6.19	6.66	+469	+7.6

Source: Champion, 1992, p. 11.

Table 4.2   Population changes from internal migration (thousands), 1990–91, by district type

	Net migration 1990–91	Change (%)
Inner London	-31	-1.2
Outer London	-21	-0.5
Principal metropolitan cities	-26	-0.7
Other metropolitan districts	-7	-0.1
Large non-metropolitan cities	-14	-0.4
Small non-metropolitan cities	-8	-0.4
Industrial districts	+7	+0.1
New Town districts	+3	+0.1
Resort, port and retirement	+18	+0.5
Urban/rural mixed	+20	+0.3
Remote urban/rural	+14	+0.6
Remote rural	+10	+0.6
Most remote rural	+36	+0.8

Source: Champion, 1998, p. 35.

people are moving to rural Britain, this numerical significance is boosted further when we note that the figures are of net migration. Given that people continue to move out of rural areas as well as come in (Weekley, 1988), the actual numbers of in-migrants is considerably higher than the net figure suggests. High gross rates of migration also enhance the potential for social change via the turnover of the population, regardless of the net figure (see Lewis et al., 1991).

The Selectivity of Counterurbanisation

It has been argued that the net migration dimension to counterurbanisation 'forms perhaps the central dynamic in the creation of any post-productivist countryside' (Halfacree and Boyle, 1998, p. 9). This point relates especially to degrees of selectiveness in the characteristics of the in-migrants to rural areas. This can be addressed first of all spatially:

- counterurbanisation migration is best described as comprising a 'cascade' (Champion and Atkins, 1996), with households often gradually moving down the population hierarchy rather than moving directly from a metropolitan to a remote rural area;
- there are strong geographical contrasts in both the origins and the destinations of migrants. Key flows include migration out of London and the other major conurbations into surrounding counties – often retaining commuting patterns into the cities – and a major net movement from the Southeast of England to remoter areas, notably the aouthwest of England and from the Midlands/northwest of England into Wales (Stillwell et al., 1996). Even relatively macro-scale flows such as these immediately warn us against over-generalising the counterurbanisation experience;
- at the more detailed level, Cloke (1985) has emphasised the importance of localised conditions in dictating the precise impact of counter-urbanisation on individual villages. Factors such as the local land market, physical environment, settlement character, housing market, employment base, cultural environment and accessibility can all influence the precise extent and character that counterurbanisation takes.

Recognising selectivity is also vitally important from a number of other angles. These tendencies reflect the findings of numerous studies of counterurbanisation within Britain since Pahl's (1965) pioneering study of three villages in Hertfordshire (a county north of London). In short, in-migrants to rural Britain tend to be biased in terms of:

- demographics: the middle aged, especially with young children, and the elderly (retirement migration); households of two or more people;
- social class: higher groups, notably households with at least one person in a professional/managerial position, with high levels of formal education;
- employment: a tendency to work in the service sector, including tourism, both via commuting and self-employment;
- ethnicity: a crucial factor, notably in the Celtic fringes of Cornwall, Scotland and Wales, is that the newcomers tend to be English. Beyond this, it should be noted that counterurbanisation is also a very white phenomenon.

Of course, from these generalisations there is a danger that the characteristics of counterurbanisers become stereotyped when, overall, a very wide range of people are involved (Bolton and Chalkley, 1989).

This latter point ties into the disparate reasons given to explain why counterurbanisation migration is taking place. The numerous explanations which have been forthcoming (see Champion, 1998, p. 33) can be split broadly into two schools: job-led versus people-led (Moseley, 1984). Summarising these very briefly, on the one hand there is migration to rural areas taking place in response to the changing space economy of Britain (see Fielding, 1982). Some of the consequences of this change for rural employment profiles are outlined below. On the other hand, people of both working age and older are 'voting with their feet' and choosing to migrate to rural areas to live. In part, this voluntarist movement can be explained with reference to perceived highly positive physical and social constructs of rural living. Indeed, for some authors the only thing which seems consistently to link the migrants together is the value that they almost all attach to such expectations (Halfacree, 1994). As we shall show later, this has clear consequences for issues of leadership, power and development in rural areas.

Socio-demographic Recomposition and Rural Poverty

Counterurbanisation processes have been a critical proximate factor in the socio-demographic recomposition of the British countryside. This comes about on the one hand from the selectivity of counterurbanisation – as discussed above – but also from the consequences of other trends impacting upon the rural population. Most fundamental are some basic economic changes. The consequences of these are reflected in the current employment structure of rural Britain, illustrated in Table 4.3.

Table 4.3   Rural employment structure, 1991 (% total)

	Britain	Urban	Rural
Agriculture, forestry, fishing	2.0	0.8	10.4
Energy, water supply, mining	4.7	4.8	4.1
Manufacturing	17.8	18.4	13.9
Construction	7.4	7.2	8.0
Services	67.3	67.9	62.4

Source: 1991 Census.

The picture presented in Table 4.3 reflects the legacy of the following (Townsend, 1991, 1993):

- an overall decline in the importance of agriculture to the rural workforce, even if it remains by far the predominant rural land-use. Numbers employed in agriculture declined from 1.33 million in 1901 to 618,000 in 1974, and stand at around half a million today;
- the impact of the so-called 'ruralisation of industry' (Fothergill et al., 1985), whereby in spite of a national decline, rural areas witnessed a net gain in manufacturing employment;
- an overall growth in the service economy, including an important role for self-employment and tourism, which tended to offset job losses in other sectors;
- the growth of female, often part-time employment, particularly in the service industries. Indeed, Townsend (1991, p. 93) concluded that 'rural areas turn out to be a leading part of Britain' for this tendency. The growth in the numbers of women in paid employment is even more impressive given the material and cultural constraints on following such a course within the rural environment (see Little, 1998).

In spite of the socio-demographic changes brought about partly through counterurbanisation and the closely linked economic transformations of the countryside, the issue of poverty and deprivation is still very much alive in rural Britain. The topic of 'rural deprivation' was forced onto the academic agenda via publications originating in campaigning charities' conferences in the late 1970s and is now a mainstay within both academic and government perspectives on the contemporary countryside (Cloke et al., 1995). However, Cloke and his colleagues feel that just regarding the problematics of rural living in terms of 'deprivation' serves to de-politicise the issue:

> deprivation has become a convenient metaphor for the rural problematic ... at once giving lip service to some notion of problems in rural life, and offering a political shrug of the shoulders about the difficulties of doing anything about it (Cloke et al., 1995, p. 353).

Thus, rural poverty needs to be reconsidered directly.

Poverty was revealed clearly in a project investigating *Lifestyles in Rural England* through interviews with 3,000 households across 12 rural areas (Cloke et al., 1994) and in a similar study in rural Wales (Cloke et al., 1997). In the English study, the overall levels of poverty found were as follows:

- 51 per cent households in receipt of less than 80 per cent of the mean national income;
- 41 per cent households in receipt of less than 80 per cent of the median national income;
- 23 per cent households in receipt of less than 140 per cent of the income support entitlement (state benefits for the poor).

These figures reflect a number of factors, including the generally low levels of wages in rural areas and residual pockets of unemployment. Clearly, then, the 'health' of rural areas suggested by patterns of counterurbanisation can mislead us as to the situation of many rural inhabitants.

However, although reclaiming the idea of rural poverty and the obvious negative material consequence which stem from this, Cloke and his colleagues also emphasised the significance of more cultural forms of exclusion within rural areas. In part this can itself result from poverty and a consequent inability to get involved in the full richness of community life. However, it also refers to issues of stigmatisation and marginalisation informed by notions of the 'correct' way to live in rural areas. Particular groups, such as one-parent families, Gypsies and black people, are often seen as somehow not 'belonging' within a rural environment by many of their neighbours. Many in-migrants, too, are not regarded as having the suitable 'cultural competence' (Cloke et al., 1998) to live in the British countryside – for example, English migrants in rural Wales (Cloke et al., 1995). However, their often strong political and economic resource base enables them to resist processes of exclusion. This resistance may be more effective than that put up by socially marginal groups, but it adds to the tensions and difficulties facing rural areas in terms of promoting effective and democratic leadership and development strategies and processes.

The Administrative Structure of Rural Britain

We now turn to consider the administrative structures governing the management of rural areas in Britain. An outline of administrative structures is helpful because it provides an understanding of the framework in which rural policy and rural politics are played out.

Central Government

The United Kingdom (UK) (59 million population, 1997) consists of four basic (national) units: England (population 49.3 million), Scotland (population 5.1 million), Wales (population 3.9 million) and Northern Ireland (population 1.7 million). Great Britain comprises just the first three nations. The predominance of England in population, if less so in areal terms, is clear from these figures.

Although the UK is a centralised state, it underwent a process of decentralisation in 1999 through the election of a parliament in Scotland with tax-varying powers and Assemblies in Wales and Northern Ireland. These new bodies are leading to a devolution of some of the authority of the central state – based in Westminster – within these countries. Nonetheless, Wales, Scotland and Northern Ireland will remain administered in part by respective government departments – operating through the Welsh, Scottish and Northern Ireland Offices – responsible for the range of policies and programmes in that country. In England, a number of separate government departments have responsibility for home affairs, environment, transport, regional affairs, defence, industry, health, social security, education and employment. The work of these departments may also cover affairs in Wales and Scotland. There are also 10 Government Regional Offices responsible for regional, urban and rural development. These offices are significant for rural areas, because it is these that administered the former Objective 5b and other European Union Structural Funds. They also play a role in the administration of national and regional rural development funding.

From 1 April 1999, nine new Regional Development Agencies in England (North East, North West, Yorkshire and the Humber, West Midlands, East

Midlands, Eastern, South West, South East, London) were given the responsibility for developing strategies for regional regeneration. They have five objectives:

- economic development and social and physical regeneration;
- business support, investment and competitiveness;
- enhancing skills;
- promoting employment;
- sustainable development.

There are some concerns that rural development issues will become hidden beneath more dominant and well-publicised urban redevelopment concerns. Specifically, there are fears that the Regional Development Agencies will concentrate on more visible, traditional urban problems such as poverty and unemployment, and will overlook or disregard the existence of such problems in rural areas. Although one representative on the board of each Agency has a rural background and experience, the dominance of urban representatives is clear (Morphet, 1998). However, for some regions (such as the northeast), even this small element of regional devolution has been welcomed for the limited degree of autonomy in economic and regional development planning that it brings.

Local Government

Excluding London and the metropolitan areas, England is governed at the sub-regional level by either County and District Councils or Unitary Authorities.[3] County Councils (34 in all) provide education and social services, strategic land use and economic development planning guidance, transport planning, highways, traffic regulation and consumer protection. Counties are subdivided into District (or Borough) Councils which provide services such as environmental health, housing, local land use planning and waste disposal (there are 238 in all). Where the two-tier structure has been superseded by Unitary Authorities, these provide both functions (there are 27 in total). Scotland comprises 32 Unitary Councils, responsible for all services, Wales has 22 and Northern Ireland has 26. The everyday work of councils is carried out by appointed officers.

All types of local government consist of elected members who have decision-making powers. Members are usually (but not inevitably) allied to one of the major political parties (Labour, Conservative, Liberal Democrat,

Green Party, plus the nationalist parties of the Scottish National Party and Plaid Cymru (Party of Wales); Northern Ireland has a very different political structure broadly divided between Unionists and Nationalists). In addition, members can sit as Independent with no stated political allegiance – this is more common in rural than urban areas, and such members are more likely to be socially and politically conservative rather than radical or left. Individual Councillors can have significant local power and influence.

Local government is significant for rural development because of its central role in land use and economic development planning. In particular, County and Unitary Councils have a direct and strategic role in economic development. There is, unsurprisingly, great variety between Councils in terms of the ways rural development issues are perceived and the manner in which policies are developed and pursued to tackle these needs. This variety is matched by diversity across rural Britain with regard to the ways in which rural development agendas devised at a local level are responded to by local elites and social groups.

Parish and Community Councils

At the most local level, Parish Councils (sometimes called parish meetings) provide and manage local facilities such as village halls and parish allotments (very small horticultural plots). There are over 8,000 Parish Councils in England. In Scotland and Wales the equivalent to the Parish Council is the Community Council. There are around 1,000 of these in Scotland and 730 in Wales. Parish and Community Councillors also act as local representatives for some District Council functions.

Parish Councils have relatively few (if any) direct powers in terms of initiating or promoting rural development activities, but they can be significant locations for debate on local development issues (Tewdwr-Jones, 1998). As the lowest tier of government, Parish Councils are important fora for the discussion of purely local (parish or village) issues; for example, provision of local recreation facilities or discussion of plans for physical developments such as housing. Members are elected, but participation in Parish Council elections can be extremely low. Crucially, established elites (for example, major landowners and farmers) often have disproportionate influence at a local level through their membership and dominance of these councils.

Finally, Rural Community Councils in England operate at County level to provide support to Parish Councils and voluntary groups. Rural Community Councils have been significant actors in different rural development programmes. As a group they are represented by the charity Action with Communities in Rural England (ACRE).

European and National Programmes for Rural Development

In this section, we consider a selection of policy mechanisms and programmes which have been central in shaping economic and social development in rural Britain. These programmes have had a pronounced impact on the abilities of different social groups to facilitate development in rural areas and, in turn, this has influenced the distribution of power and the role of different elites in rural areas. The programmes outlined below have been selected on the basis of the specific points about leadership and local power that can be drawn from each.

Objective 5b of the European Structural Funds

Eleven Objective 5b regions – 'disadvantaged rural areas' – were designated in Britain for the period 1994–99. One of these was in Wales (Rural Wales), four were in Scotland (Dumfries and Galloway, Rural Stirling and Upland Tayside, Grampian, Borders) and six were in England (East Anglia, South West, Northern Uplands, English Marches, Lincolnshire, Midlands Uplands). A total of ECU 819 million (£682 million) was granted for rural economic development in these regions for a total population of 4.5 million people (see Ward and McNicholas, 1997b).

Objective 5b engages with issues of local power and leadership in a number of ways. First, the very designation of Objective 5b has promoted what some have termed the 'Europeanisation' of policy networks (Ward and Woodward, 1998), in that it has promoted greater awareness of the European context in which uneven regional development takes place amongst those with a responsibility for rural development at a local level. Awareness ranges from increased knowledge amongst local council officers about the mechanisms by which the European Commission operates in promoting rural development, to increased conceptual understanding of the nature of uneven rural development in Europe.

Second, availability of Objective 5b funds has been important in establishing rural development problems as a priority for policy intervention at a regional and national level in Britain. The designation of areas for receipt of funding has highlighted the existence of pockets of economic underdevelopment within broadly prosperous areas, and has signalled the persistence of economic underdevelopment in lagging regions of the UK.

Third, the process of bidding for Objective 5b status has enabled some regional government authorities to define a locality, its problems and its aspirations. Single Programming Documents, outlining the characteristics of the area for which Objective 5b funding is sought and setting out the objectives for the proposed actions, have in the main drawn different actors together in the process. This process requires areas to 'think themselves into being' and has helped actors in localities to define an area and its potential. Of course, as part of this process, the views of some social groups (often those most closely connected within existing power structures) have taken precedence over others.

Fourth, the process of defining an area, its problems and possible solutions has also helped encourage rural communities at the local level to consider economic development solutions to locally-defined problems. In particular, the process of application for Objective 5b funds, and the process in some areas of conducting Village Appraisals of local needs, has raised new questions for rural communities about local needs (Ward and McNicholas, 1997a). Again, one outcome of this process has been that (inevitably) the perceptions of a locality, its problems and its prospects that dominate are often those of local elites, frequently with established interests in existing power structures.

Lastly, designation of areas for the receipt of Objective 5b funding has been important in the process of defining local elites, through the incorporation of individuals within administrative groups responsible for controlling the funds, and through the funding of individual projects. Some individuals and groups have gained political capital out of the processes of defining the local programme and determining the expenditure of funds. There is currently some debate in rural policy circles about the degree to which local actors (be they elected Councillors, local council Officers or more informal community representatives) have control over the trajectory of Objective 5b. This is an important issue in consideration of local power and leadership. Also, there have been indications of a 'democratic deficit' in the administration of Objective 5b funds, with decisions on the distribution of funding taken by appointed officials rather than by elected members. Furthermore, in some areas there have been criticisms of complexity, bureaucracy and delays in decision-making, which have further reduced the overall accountability of the programme. Objective 5b projects have not generally been perceived as being as 'locally owned' as those of LEADER (see below) (Ward and Woodward, 1998).

Under the new Rural Development Regulation, set out under Agenda 2000 and the reform of the Structural Funds, from late 2000 a new Objective 2 designation is available. Lessons learnt from the Objective 5b (and Objective 1) programme will, it is hoped, feed into the Objective 2 programme. Nonetheless, fears have been voiced that Objective 2, a designation driven by unemployment levels, will concentrate on urban and regional development, marginalising the needs of rural areas.

LEADER

LEADER (Liaisons Entre Actions de Développment de l'Economie Rurale) is a European Union initiative in operation, to date, for the periods 1989–93 (LEADER I) and 1994–99 (LEADER II). In the UK, 66 LEADER II territories were designated (20 in England, eight in Wales, 14 in Scotland, 24 in Northern Ireland). LEADER is highly significant for rural development and for issues of local leadership and power. Although very small amounts of money have been committed (c. £14 per capita in LEADER II areas, compared with c. £200 per capita in Objective 5b areas in UK: Ray, 1998), the programme has been based on the premise of endogenous development. The understanding here is that people in LEADER territories should participate in the design of development activity with themselves as beneficiaries. For some, LEADER's ethos is highly democratic; for others it is perhaps anarchic, although the EU's DGVI still maintains a high level of control.

LEADER creates new spaces for rural development. LEADER territories are newly inscribed and do not conform to local authority boundaries which, 'in the conventional view of representative democracy, are the legitimate areas for representing the aspirations' of people and in which to spend public money to address those aspirations' (Ray, 1998, p. 25). For this reason, some commentators see LEADER as having great potential to enable participative democracy:

> Marginalised rural economies are enabled to restructure themselves so as either to participate more effectively in national/European/global economies or to resist largely extra-local forces that have produced local vulnerability and decline. By re-drawing the boundaries of policy intervention and by thus valorising local resources, local control of, and participation in, social and economic development can theoretically be enabled (Ray, 1998, p. 25).

As part of the process of creating new spaces for rural development, LEADER can enable some social groups to take part more fully in local economic and social development initiatives. Thus, some authors suggest that

LEADER is essentially empowering to local groups (see Lowe et al., 1998). Furthermore, pockets of social exclusion may become more visible and may be enabled to help themselves (Ray, 1998). In addition, LEADER areas can act as a prism through which different agendas with a rural dimension can ground themselves in localities; for example, those of environmental organisations. LEADER opens up possibilities for different voices to be heard.

However, there have been debates as to whether LEADER really encourages democratic decision-making at the local level. Ray (1998) notes the debate in Scotland that it is in fact anti-democratic because much of the initiative for local LEADER groups has come from the private sector. 'Ownership' of the idea and the local programme has remained within this sector, to the detriment of open and democratic decision-making. It is certainly the case that local LEADER groups are self-appointed, raising issues of accountability and legitimacy in their activities. The consequences of this for issues of local leadership and power are obvious. From the opposite direction, there have also been criticisms about the lack of control given to LEADER groups by central/regional government, despite the responsibility for the success of local programmes resting firmly with local actors. Ultimately, there are issues about the extent to which power really does reside in localities with LEADER.

Overall, there has been a great variety in the effectiveness of LEADER groups in promoting rural development. There have been both great successes and great failures. Areas with less success have blamed high local expectations for their LEADER projects, which have been impossible to match in reality.

Rural Development Programmes

Rural Development Programmes (from April 1999 the responsibility of Regional Development Agencies) are programmes for the expenditure of grant funding for rural economic and social development. There are 31 Rural Development Programmes in England, covering 35 per cent of the land area and 6 per cent of the population (2.75 million people).[4] Around £10.5 million was spent in England in 1997–98. The amount of funding is small but Rural Development Programmes can be an important source of seed-corn or matched funds.

Issues of local power and leadership appear in a number of aspects of Rural Development Programmes. First, they encourage the use of partnerships between various institutions in order to facilitate, manage and often part-fund rural development initiatives. Partners can include training bodies, business support bodies, tourist boards, community councils and regeneration agencies. As with LEADER, whilst the notion of partnership is often felt (politically at least) to be an attractive one, there are often issues of the balance or share of power between bodies involved in development partnerships. For this reason, many are sceptical of the rhetoric of partnership and have concerns about the uneven distribution of local power that unequal partnerships can produce.

Second, Rural Development Programmes entail the devolution of decision-making to the level of the County. Devolved decision-making, including the ability of localities to determine local needs and solutions is generally thought to be an effective and accountable mechanism for the delivery of rural development funding. Inevitably, the criticism can be made that the views of some local groups and elites are heard in preference to others.

Finally, Rural Development Programmes promote an integrated approach to rural development, involving consideration of social, economic and environmental factors. This helps to overcome the overlaps and even conflicting aims and objectives that more narrowly sectoral responsibilities can produce.

Rural Challenge

The Rural Challenge programme was initiated in 1994 by the Rural Development Commission, a government agency for economic and social regeneration in rural areas of England. Bidding to this funding programme was undertaken by individual projects in four rounds until 1997. In total, 27 projects were awarded up to £1 million each. Match funding was also to be sought from other sources. One important factor in determining whether bids gained money was whether they strengthened the capacity of local communities to drive regeneration. The programme has now finished, but is worth considering here because of the contribution that it has made to debates on the most appropriate format for rural development programmes. Themes raised in the previous three subsections are reiterated.

First, Rural Challenge was based on the principle of partnership between community groups, non-governmental organisations, government agencies and the private sector. Indeed, the establishment of partnerships in support of specific bids for funds was a precondition for eligibility. Whilst the rhetoric of partnership was attractive, particularly within government, the problems of inequalities between partners sometimes reduced the capacity of local communities to benefit fully from the funding provided by the programme (see Little et al., 1998).

Second, bids and projects demanded a great deal of input from local communities. Community groups had to be extremely professional in order to obtain funding, which could work against the commonplace ethos of informal methods of political operation. Consequently, established community organisations had a better chance of succeeding in obtaining funding from Rural Challenge; local political elites remained highly significant (see also Rural Development Commission, 1999). Yet, the process of obtaining funding was important in some communities in terms of capacity building and establishing mechanisms for self-help.

Third, there is some evidence that a limited range of social groups were involved in Rural Challenge projects. Newer rural residents ('incomers'), who (as shown in Section 1) were more likely to have professional qualifications and a wider range of organisational skills, were dominant on Rural Challenge boards. In some areas, it was very difficult to get key sectors of the community involved; established political interests remained just that.

In conclusion, we have argued that policy mechanisms and programmes have shaped local power relations in significant and not always socially equal ways. Moreover, it should be remembered that such programmes operate unevenly across rural Britain, which in itself is highly differentiated. This differentiation is compounded by local cultural difference, to which we turn in the next sections.

Rural Discourses in National Policy-making

Following the contextual and policy introductions, this next section shifts the focus to issues surrounding the formulation and implementation of rural development policy at the national level. We examine how scholarly inquiry has singled out discourses of rurality as an object for scrutiny, and draw on this body of literature to assess how rural development agendas are established, how rurality is conceptualised within rural policy discourses, and how emerging policy networks and social movements have coalesced around rural development policy and practice. We do not suggest that this discursive turn has somehow been responsible for shifts in rural policy; rather, that the study of rural policy can benefit from insights obtained through examination of the discursive content and function of policy statements.

Redefining the Rural

Over the last decade or so, rural studies in Britain, in common with most other areas of social scientific endeavour, has taken something of a 'cultural turn'. For rural studies, the shift has revolved around both a consideration and a reconceptualisation of the meaning of 'rural' and some of the implications which flow from this. As Cloke (1997, p. 368) puts it:

> Within rural studies this [cultural] turn has been strongly represented, not least in contributions which have explored the ways in which the signs and significations of rurality have been freed from their referential moorings in geographical space.

This freeing-up of the rural comes from seeing rurality defined as a socially constructed and discursive category rather than as some form of *a priori* material reality, there simply to be discovered and mapped out on the ground. Thus, from a postmodern perspective, 'the desire to hold the rural and urban as separate' is regarded by Murdoch and Pratt (1993, p. 415) as a 'modernist impulse'.

Taking this idea further, Halfacree (1993) has recommended that the rural can be defined either as a locality, where significant societal structures operate within a distinctive space on the ground, or as a social representation of space. The latter perspective comes from acknowledging how 'the rural' and its synonyms are words and concepts understood and used by people in everyday talk' (Halfacree, 1993, p. 29; see also Jones, 1995). From such an acknowledgement, the rural as a social representation comprises:

> an organisational mental construct which constitutes what is 'visible' and must be responded to, relates appearances and reality and even defines reality itself … the rural consists of both abstract concepts and concrete images (Halfacree, 1995, p. 2).

Developing this idea, Murdoch and Pratt (1993, p. 425) coin the term 'post-rural' to 'highlight the reflexive deployment of 'the rural'', possibly disconnected from any recognisable space on the ground at all (see also Pratt, 1996).

The notion of the rural as an imaginative category rather than material reality has proved popular in rural studies. Its definition and contours become key issues of culture, ideology and power. For our purposes here, the insights drawn from academic considerations of this reconceptualised rurality are useful in showing how issues of local power and leadership are shaped by cultural processes, in particular.

The Rural Idyll and Processes of Restructuring

Examining the character of the rural as defined by a social representation of space, Halfacree (1994, 1995) has argued that one family of representations – the 'rural idyll' – is of fundamental importance. This representation can be described generally as:

> physically consisting of small villages joined by narrow lanes and nestling amongst a patchwork of small fields where contented ...cows lazily graze away the day. Socially, this is a tranquil landscape of timeless stability and community, where people know not just their next door neighbours but everyone else in the village (Halfacree and Boyle, 1998, pp. 9–10).

Within such a countryside:

> is pictured a less-hurried lifestyle where people follow the seasons rather than the stock market, where they have more time for one another and exist in more organic community where people have a place and an authentic role (Short, 1991, p. 34).

At its most crude and basic, as these outlines suggest, the idea of the rural idyll can be summarised in clichés (often visual) of the English village, with thatched cottages, neatly hedged fields, old farm buildings and so on.[5] At a more complex level, the idea of the rural idyll goes beyond the visual aesthetic. It can be seen as a powerful element within English cultural life, shaping dominant ideologies about the countryside, what it is, and (most crucially) what it should be. The social construction of rurality as the rural idyll is powerful, and has been linked to the socio-demographic processes of counterurbanisation and to other processes of restructuring in the countryside. For example, Murdoch and Marsden (1994) show the connections between culture and political economy visible in the restructuring of the countryside of Buckinghamshire, northwest of London.

The rural idyll further ties into issues of leadership, local power and restructuring in that it presents a culturally dominant or hegemonic ideal for the practice of rural social relations. The idyll also exerts normative pressure on the workings of authority within rural society. This dominant cultural norm endorses, for example, established ideas of social structure based on hierarchies, which in turn reflect the hegemony of landed interests and agriculture (and, until recently, the church). Notions of the rural idyll are powerful also in that they deny the existence of problems such as poverty and deprivation in the countryside; if everything looks perfect and if a particular social order is naturalised in this representation then the existence of problems can be readily denied (Woodward, 1996). Even where problems are acknowledged, the prevalence of idyllised notions locates their source as being somehow 'outside' rural society. The ideology of the rural idyll is, in part, about maintaining an established social order. Established political elites and the interests of key social groups are prioritised, dissent is erased.

Discourses of Rurality and Rural Policy Statements

Discourses of rurality are important because they provide a vital part of the context in which policy statements are developed. Their influences often work in subtle or at least taken-for-granted ways. Rural policy in Britain was effectively, since 1945, synonymous with agricultural policy, with an assumed benign leadership role assigned to the agricultural community. This led, for example, to 'agricultural exceptionalism' (Newby, 1987, p. 216) within planning i.e. a near exemption of agriculture from planning regulations. Policies affecting social and economic development in rural areas operated without reference to any over-arching policy vision beyond the heavily idyllised assumptions encapsulated by the Scott Report of 1942 and enshrined in key post-war legislation, notably the Agriculture Act (1947), the Town and Country Planning Act (1947) and the National Parks and Access to the Countryside Act (1949). Even in the 1980s and early 1990s, with the emerging trends of post-productivism and counterurbanisation, little had changed.

In 1995, official White Papers (published individually for England, Scotland and Wales) rectified this situation a little by providing an account of how the Conservative government saw the future social, economic and community development possibilities for rural areas. Two key points come from these White Papers. First, there was an emphasis on re-energizing the engagement of local citizens and institutions in the development process, getting them to assume responsibility for elements of the government of their own communities – 'governing through communities' (Murdoch, 1997). Local communities should identify and cater for their own unique needs. Murdoch argues that this is in part a mechanism for the withdrawal of the state from areas of rural need. The White Papers' argument presupposes a particular vision of rural areas as possessing already viable and coherent communities (geographical or interest based) who can proceed with this vision of 'self-government' and self-promotion. The idea of 'community' is, of course, central to notions of the rural idyll (see Murdoch and Day, 1998).

Second, the English White Paper is a testament to specific English preoccupations with landscape and wildlife; it contains little on agricultural pollution, for example (Hodge, 1996). This limited vision is extended more generally in a focus on small, individual initiatives rather than providing a more thorough analysis of the underlying forces for rural change or stasis. Again, we can draw connections between how these policy statements defer to a specific vision of the rural and rurality.

Government seems to have acquired a taste for outlining its vision of the countryside and a new Rural White Paper for England was produced in November 2000 by the New Labour government. This was preceded by the Performance and Innovation Unit's (1999) report on *Rural Economies* suggestion of a more sweeping and comprehensive review of rural policies, urging 'modernisation'. Deployment of this idea might yet offer a radical break with past visions but this is certainly far from guaranteed. Dominant discourses of the rural still seem to hold powerful sway over the policy making environment.

New Rural Governance?

The potential challenge to the hegemonic role of the discourse of the rural idyll goes hand-in-hand with other changes in the rural power agenda. The term 'governance' is used to signify new (to the mid-1990s) configurations of public, private and voluntary sector involvement in the government of communities and territories. Probably most significant is an emphasis on partnership, not least the encouragement of partnerships between public, private and voluntary sectors in order to promote development. As we outlined above, this has been a very strong feature of recent rural development policy. Analysts have taken this move towards partnership as indicative of a move towards new styles of governance in rural areas (Marsden and Murdoch, 1998, even a 'new rural governance'. A significant point here concerns how new modes of governance might affect how development agendas are set, how rurality is interpreted and how it influences such agendas, and how different policy networks and social movements hold influence over rural development policy (Ward and McNicholas, 1998).

The development agendas which are set within the new rural governance suggest the influence of more business-orientated commodified notions of the rural than that inscribed by the rural idyll. In line with New Labour politics more generally, emphasis is given to:

- what is possible to achieve within the scope of clear time-limited programmes; pragmatism over broader visions;
- the desires of the funding bodies, despite giving lip-service to ideas of endogenous development. Although bottom-up development is accepted as desirable, agendas promoting broadly economic development at the expense of environmental or social objectives are dominant in practice (Wood, 1998);
- the principle of partnership, when in fact partnerships might be ineffective, unbalanced or weak, as noted earlier.

This new rural governance is not, however, an uncontested domain. For example, some social movements/groups (whether organised or not) have managed to influence development policy to make it more responsive to local needs. Examples include the Rural Community Councils, who can be powerful if they have a strategic vision. This is one of the aims of their umbrella organisation, ACRE. Likewise, the Rural Housing Trust, with its slogan of 'Village Homes for Village People' has since the late 1980s set up a number of charitable associations which aim to build small groups of houses, flats and cottages to address local needs (Williams and Bell, 1992).

Rural Restructuring, Power and Leadership at Regional and Local Levels

The roles of different groups of actors at the local level are crucial to the rural development process. These roles should be understood in the context of rural discourses and their potential mediation of and incorporation into policy, as discussed above.

Old and New Elites

Processes of social recomposition, taking place as part of wider processes of rural restructuring, have produced marked changes in the social structure of rural areas and in the opportunities for political participation amongst those in the countryside. Most significant has been the decline in authority of the old landed gentry. Older social relations rooted in paternalism have to a great extent given way to new political formations based around new elites drawn primarily from the ranks of mostly middle class in-migrants to rural areas. Members of the landed gentry still provide public comment on rural issues, such as deprivation, not least in the House of Lords, but their voice comes across as increasingly anachronistic.

Some old elites (or, at least, prominent social groups) have become more vocal in commentating on contemporary rural issues. For example, the Women's Institute, traditionally a bastion of middle-class values and respectability (Hughes, 1997), has taken a keen interest in local economic development issues as part of a wider recognition of the significance of female involvement in economic activity.[6] In another example, the Church of England has shown a long-standing interest in issues of poverty and deprivation in rural areas, and has prompted national debates on rural futures to consider the excluded, the poor and the marginalised. A revitalisation of old elites was also apparent in the formation of the Countryside Alliance in 1997, a group campaigning primarily for the right to hunt animals with dogs, pursuits which are presently under severe threat from Parliament.

In terms of new elites, processes of counterurbanisation have brought new people into rural areas, often with their own ideas on the most appropriate forms that development (if any) should take. Generally, the new rural residents are resistant to development, as it impinges on their notion of what is appropriate for 'rural' areas. Where there is desire for change, however, new elites have been instrumental, through their highly informed participation in the planning process (see Abram et al., 1996), in moulding villages and village development into the image of the rural idyll, for example (see Murdoch and Marsden, 1994). Looking slightly differently at the basis of action for these new elites, Phillips (1998a, 1998b) has argued that social categorisations based on moral codes are increasingly important. The moral values underpinning people's own perceptions of their and others' class and social position are markers of different political perspectives and attendant activities (see also Bell, 1994).

New insecurities have accompanied the arrival of new elites. This reveals something of the contradictory character of contemporary rural life:

> While the perceived tranquillity, changelessness and security offered by rural areas continue to attract migrants from towns and cities, new insecurities for some social groups are generated as a result (Ward and Lowe, 1998a, p. 3).

In short, newcomers can bring social and political values which may be destabilising for some older established rural residents. Nowhere is this clearer than on the Celtic fringes, where English incomers are seen as a new elite group who are a disruptive influence on established tradition and rural life (for example, see Cloke et al., 1995).

Such changes have also been noted by Abram (1999), who argues that, increasingly, social action in rural areas is just as likely to be organised around other interests and identities than simply those associated with class – for example, religious, environmental, political, generational or gender identities. Again, new social formations are potentially destabilising to old elites and older social orders where class identities much more directly determined participation in social action and economic and social development.

Insiders and Outsiders in Rural Development Strategies

Social recomposition in rural Britain is also shifting the profile of those directly involved in rural development strategies and those who are left out. This shift is strongly tied in with the emergence of the new elites, described above, who often 'capture' the development process, but it also has a managerialist dimension. For example, Ray (1999a) notes the importance of the locality-based professional, employed to implement the programmes for rural development. Such individuals play a crucial role in translating a policy on paper into action on the ground. Specifically, these individuals can have an 'interpretative' influence, in that they can shape the ways in which development projects evolve at the local level because of their individual perceptions of the role of such programmes, the way they might match them with local needs, and the influence of the individual's beliefs and biography.

The power of elites and the policy professionals raises concerns for the socially excluded – those who may not have much voice in the determination of rural development strategies. These groups may include former elites but, more significantly, there remain many within rural society who have never been empowered. Despite all the talk of the benefits of endogenous development, there are still many structural barriers to full empowerment within rural communities. Many of those with most to benefit from rural development programmes may not necessarily have access to such programmes in practice (see Ellis, forthcoming). Moreover, this continued exclusion is not just a question of class. Neither is it reducible to the 'neglected' groups outlined below. For example, in order successfully to access much development funding the applicant needs to be 'on message'. They must be able to reflexively deploy the current development/regeneration vocabularies promoted by government. At the end of the year 2000 this means speaking the language of New Labour, notably the discourse of partnership, stake-holding and proactive communities.

Regionalism

Rural development has assumed a clear regional dimension in some areas. For example, in England there is a strongly southeast-centred debate on new housebuilding, with arguments over the location for the (disputed) 1.4 million new houses 'needed' for the region by 2016 (*Guardian*, 1999). Should these houses be generally dispersed around the region, should they be focused into selected settlements (including possible new towns), or should they be concentrated onto already developed but now derelict 'brownfield' sites in the cities? The debate and its eventual outcome has and will continue to galvanise social groups – both old and new – empower them differentially, and reflect these differential power relations. Former political adversaries can make common cause, as in the alliance between conservatives and radicals within the Campaign Against Stevenage Expansion (CASE, 2001). Alternative discourses of rurality, too, will feature strongly within these processes (see Abram et al., 1996, 1998).

The devolution accorded to Scotland and Wales in 1999 has also stimulated regional debates around development, rural change, and local power and leadership. For example, one of the first tasks of the Scottish parliament was to discuss the controversial issue of land rights and ownership. In 1997, the Scottish Office established a Land Reform Policy Group to consider various dimensions of the 'land question' (Lloyd and Danson, 1998, 1999). Drawing on the Group's 'Recommendations for action' (Land Reform Policy Group, 1999), plans for a change to the Scottish landownership system were announced as an initial task for the parliament in 1999. This priority seems to be still alive today, with the publication of proposals for legislation by the new Scottish Executive (1999). Reforms of the remnants of feudalism will promote community buy-out of estates which come onto the market. Once again, discourses of rurality become drawn into this debate, such as the Scottish crofters' aims of creating a more populated and ecologically sustainable countryside (Halfacree, 2000).

Gender

Much attention has been paid in the literature to the role of women in maintaining established social (and economic – see note 6) relations in rural areas. Little and Austin (1996), for example, have examined how social and community life in the countryside is usually dominated by activities undertaken by women. Women are more likely to predominate in unpaid and voluntary positions within a community organisations (such as the church) that provide an important focus for rural communities (Seymour and Short, 1994). This is partly because of lower levels of female engagement with the labour market, giving women greater opportunities to facilitate social activities, but also because of the pressure of social expectations about the 'natural' role or women and their place as carers and nurturers in the community (Little and Austin, 1996). Hughes (1997) notes how social expectations about the 'natural' role of women can be oppressive, with participation in community and social life regarded as a duty rather than a joy. Whilst reluctance to participate in social life can compound isolation, inclusion can often reinforce gender relations embedded within patriarchal values which operate against women's equality interests.

Gender relations within rural areas are far from static, however, although changes are not always seen in a positive light. For example, fieldwork for the *Lifestyles in Rural England* project (Cloke et al., 1994) found many interviewees highlighting the changing position of women in the labour market, as well as wider processes of economic restructuring, as being responsible for the disruption of an established social order caricatured within the ideology of the rural idyll.

Given their significance, we need to ask whether or not the voices of women are heard in rural development debates. There are various responses to this question. However, broad socioeconomic data suggest that, nationally, women as a group have much to benefit from rural development simply in terms of improving their unequal access to the labour market, childcare, transport and healthcare. Such improvement would be relative to both rural men and to urban women. Furthermore, locally-derived information suggests that innovative rural development projects can have added value when targeted at women. For example, increasing female economic activity tends to bring down poverty at the household level, with knock-on effects on child health, education, etc.

Other Groups

As we have seen above in the case of gender, the idea of the rural idyll is highly selective, as is the fate of any singular discourse of the rural which fails to accord a space for inherent diversity. However, the increasing differentiation present within the British countryside means that any hegemonic status for this rural idyll is under ever-increasing challenge. Whilst we consider that the ideology of the rural idyll remains culturally dominant, alternative social constructions of rurality also exist. As Philo (1992) notes, these 'neglected rural geographies' require recognition, elaboration and investigation, not least because of the potential they indicate for change within established social orders. Different social groups will have varied perceptions of both the rural development process, and of their position in relation to it (i.e. aspirations to change, for themselves or the system).

Examples of groups and experiences covered by the neglected rural geographies label are widespread. They include considerations of ethnicity and constructions of the countryside (Agyeman and Spooner, 1997; Kinsman, 1997); gay and lesbian experiences of rurality (Bell and Valentine, 1995); 'New Age travellers' and other alternative lifestyle cultures (Halfacree, 1996; Sibley, 1997); children's lives in the countryside (Jones, 1997); indeed, the experiences of all excluded groups (Shucksmith and Chapman, 1998; Cloke and Little, 1997; Milbourne, 1997). Within these alternative experiences of rurality come different conceptualisations as to what the countryside is and, perhaps most crucially, what it should be. When mobilised, these alternative ruralities can have significant consequences for rural development, as they challenge established regimes of leadership and local power. Such mobilisation is, however, generally unrealised.

Conclusion: Rural Development and Empowerment

We began this chapter by showing how Britain's rural areas have been and still are experiencing profound socio-demographic and economic restructuring, driven not least by migration processes. From this perspective alone, the sheer diversity present today merits reiteration. This diversity is enhanced by a considerable administrative complexity, which itself is also very much in flux. In summary, we have what Marsden et al. (1993) describe as a highly 'differentiated countryside' for the twenty-first century.

All of this complexity must be acknowledged from the beginning in any consideration of issues of power and restructuring in rural areas. Factors influencing economic development trajectories and/or local political processes will vary greatly. For example, contrast the crofting areas of northern Scotland and the intensive agri-business dominated agricultural areas of East Anglia in eastern England. Equally, the economic and social pressures on, for example, the protected valorised rural farming landscapes of the Cotswolds in central-southern England will be very different from those affecting the ex-mining ex-industrial countryside of northeastern England. With this contextual backdrop in mind, we end this chapter with a number of still open questions with respect to issues of leadership and local power in rural Britain.

- *Who controls, manages and benefits from rural development programmes?* Control is uneven, varying with factors such as the distribution of power at different administrative levels, the views on development held at each level, who sits on the committees and draws up the policies, how committed management committees are to soliciting opinion from local communities and translating their perceived requirements into actions. Pulling out and detailing this unevenness is a key challenge for future research.
- *Has there been a redistribution of power through the various development programmes?* Ideas of endogenous development, in particular, have been incorporated into development programmes (for example, Objective 5b, LEADER). They certainly 'stir up' the local development issue, but it often seems that ingrained power relations still hold, or new elites are facilitated in their capturing of the development process. For example, Objective 5b programmes have not generally promoted the levels of economic development required to constitute true 'structural adjustment', part of which would entail a reconfiguration of local power relations, although there is some local evidence for small-scale changes which enable empowerment for specific groups, communities or localities. Crucially, Ray (1999b) points out that current mechanisms for evaluating rural development programmes undermine the radical potential of endogenous development initiatives by focusing on end products rather than on the processes and structures put in place by these programmes. McNicholas and Woodward (1999) conclude that greater strategic thinking is needed for future programmes, with an acceptance that it is the lasting impacts of such programmes (such as local capacity building) which counts.[7]
- *How can different sets of actors in the redevelopment process, at both regional and local levels, use their bargaining power to best advantage within the context of national and European rural development programmes?* This challenge ultimately involves these groups of actors taking control of the rural development process. This is difficult, as it requires a certain capacity, a knowledge about how systems work, and an acquiescence from those in regional and national government that local control is desirable. Even if they succeed, this can pose problems in terms

of issues such as legitimacy, accountability and democracy. This is especially the case when specific discourses of the rural, incorporating often quite sectional interests, become hegemonic within the development process.

Overall, in conclusion, issues of empowerment, and the complex political, social and cultural issues contained therein, appear to be coming ever more significant within the development agenda of rural Britain.

Notes

1   Unless otherwise qualified, the term 'rural' is used here to denote areas outside of settlements of 10,000 people or more (a definition deployed by government). This contrasts with definitions of 'rural' deployed in other European contexts. There has been significant and on-going debate amongst rural geographers and sociologists in Britain over the past two decades about the precise meaning of 'rural' and whether it has utility in social scientific debates. We return to this point later in the chapter.

2   See Boyle (1995) for details of this classification scheme and of that used in Table 4.2.

3   This mixed structure followed an official review of UK local government, begun in 1991, which saw the introduction of Unitary Authorities in some parts of the country.

4   Scotland, Wales and Northern Ireland have different administrative structures for rural development, linked most notably to Highlands and Island Enterprise, the Welsh Development Agency and Northern Ireland Executive, respectively.

5   It should be noted that the rural idyll is specifically an English social construction, rather than British, and more applicable to the south and east of England than to the north and west.

6   In rural areas, small-scale local economic enterprises run typically by women, such as on-farm holiday accommodation, often provide essential sources of income to otherwise impoverished households.

7   There are concerns that the new rural development Regulation set out under Agenda 2000 will not take account of ideas of endogenous development to the same extent (Ward and Lowe, 1998b).

References

Abram, S. (1999), 'Up the Anthropologist: Power, Subversion and Progress (Another Reply to Hoggart)', *Journal of Rural Studies*, vol. 15, pp. 119–20.

Abram, S., Murdoch, J. and Marsden, T. (1996), 'The Social Construction of 'Middle England': the Politics of Participation in Forward Planning', *Journal of Rural Studies*, vol. 12, pp. 353–64.

Abram, S., Murdoch, J. and Marsden, T. (1998), 'Planning by Numbers: Migration and Statistical Governance', in P. Boyle and K. Halfacree (eds), *Migration into Rural Areas: Theories and Issues*, Wiley, Chichester, pp. 236–51.

Agyeman, J. and Spooner, R. (1997), 'Ethnicity and the Rural Environment', in P. Cloke and J. Little (eds), *Contested Countryside Cultures: Otherness, Marginalisation and Rurality*, Routledge, London, pp. 197–217.

Bell, D. and Valentine, G. (1995), 'Queer Country: Rural Lesbian and Gay Lives', *Journal of Rural Studies*, vol. 11, pp. 113–22.

Bell, M. (1994), *Childerley*, University of Chicago Press, London.

Bolton, N. and Chalkley, B. (1989), 'Counter-urbanisation – Disposing of the Myths', *Town and Country Planning*, vol. 58, pp. 249–50.

Boyle, P. (1995), 'Rural In-migration in England and Wales 1980–1981', *Journal of Rural Studies*, vol. 11, pp. 65–78.

CASE (Campaign Against Stevenage Expansion) (2001), homepage www.case.org.uk (accessed January).

Champion, A. (ed.) (1989), *Counterurbanisation: The Changing Pace and Nature of Population Deconcentration*, Edward Arnold, London.

Champion, A. (1992), 'Introduction: Key Population Developments and their Local Impacts', in A. Champion (ed.), *Population Matters: The Local Dimension*, Paul Chapman Publishing, London, pp. 1–21.

Champion, A. (1998), 'Studying Counterurbanisation and the Rural Population Turnaround', in P. Boyle and K. Halfacree (eds), *Migration into Rural Areas: Theories and Issues*, Wiley, Chichester, pp. 21–40.

Champion, A. and Atkins, D. (1996), *The Counterurbanisation Cascade: An Analysis of the 1991 Census Special Migration Statistics for Great Britain*, Seminar Paper 66, Department of Geography, University of Newcastle upon Tyne.

Champion, A. and Congdon, P. (1988), 'Recent Trends in Greater London's Population', *Population Trends*, vol. 53, pp. 7–17.

Cloke, P. (1985), 'Counterurbanisation: A Rural Perspective', *Geography*, vol. 70, pp. 13–23.

Cloke, P. (1997), 'Country Backwater to Virtual Village? Rural Studies and the "Cultural Turn"', *Journal of Rural Studies*, vol. 13, pp. 367–75.

Cloke, P., Goodwin, M. and Milbourne, P. (1995), '"There's So Many Strangers in the Village Now": Marginalization and Change in 1990s Welsh Rural Life-styles', *Contemporary Wales*, vol. 8, pp. 47–74.

Cloke, P., Goodwin, M. and Milbourne, P. (1997), *Rural Wales. Community and Marginalization*, University of Wales Press, Cardiff.

Cloke, P., Goodwin, M. and Milbourne, P. (1998), 'Inside Looking Out; Outside Looking In. Different Experiences of Cultural Competence in Rural Lifestyles', in P. Boyle and K. Halfacree (eds), *Migration into Rural Areas: Theories and Issues*, Wiley, Chichester, pp. 134–50.

Cloke, P., Goodwin, M., Milbourne, P. and Thomas, C. (1995), 'Deprivation, Poverty and Marginalization in Rural Lifestyles in England and Wales', *Journal of Rural Studies*, vol. 11, pp. 351–65.

Cloke, P. and Little, J. (eds) (1997), *Contested Countryside Cultures: Otherness, Marginalisation and Rurality*, Routledge, London.

Cloke, P., Milbourne, P. and Thomas, C. (1994), *Lifestyles in Rural England*, Rural Development Commission, London.

Ellis, A. (forthcoming). *Power and Participation in Grassroots Rural Development: the Case of LEADER II in Wales*, PhD thesis, University of Wales Swansea.

Fielding, A. (1982), 'Counterurbanization in Western Europe', *Progress in Planning*, vol. 17, pp. 1–52.

Fothergill, S., Gudgin, G., Kitson, M. and Monk, S. (1985), 'Rural Industrialization: Trends and Causes', in M. Healey and B. Ilbery (eds), *The Industrialization of the Countryside*, Geo Books, Norwich, pp. 147–59.

Guardian (1999), 'Prescott Urged to Approve 1.4m Homes in South-East', 10 November.

Halfacree, K. (1993), 'Locality and Social Representation: Space, Discourse and Alternative Definitions of the Rural', *Journal of Rural Studies*, vol. 9, pp. 23–37.

Halfacree, K. (1994), 'The Importance of 'the Rural' in the Constitution of Counterurbanization: Evidence from England in the 1980s', *Sociologia Ruralis*, vol. 34, pp. 164–89.

Halfacree, K. (1995), 'Talking about Rurality: Social Representations of the Rural as Expressed by Residents of Six English Parishes', *Journal of Rural Studies*, vol. 11, pp. 1–20.

Halfacree, K. (1996), 'Out of Place in the Country: Travellers and the "Rural Idyll"', *Antipode*, vol. 28, pp. 42-72.

Halfacree, K. (2000), 'Beyond Counterurbanisation: Renegotiating Rurality Through Recent "Back to the Land" Movements', unpublished manuscript.

Halfacree, K. and Boyle, P. (1998), 'Migration, Rurality and the Post-Productivist Countryside', in P. Boyle and K. Halfacree (eds), *Migration into Rural Areas: Theories and Issues*, Wiley, Chichester, pp. 1–20.

Hodge, I. (1996), 'On Penguins on Icebergs: the Rural White Paper and the Assumptions of Rural Policy', *Journal of Rural Studies*, vol. 12, pp. 331–7.

Hughes, A. (1997), 'Rurality and "Cultures of Womanhood". Domestic Identities and Moral Order in Village Life', in P. Cloke, and J. Little (eds), *Contested Countryside Cultures: Otherness, Marginalisation and Rurality*, Routledge, London, pp. 123–37.

Jones, O. (1995), 'Lay Discourses of the Rural: Developments and Implications for Rural Studies', *Journal of Rural Studies*, vol. 11, pp. 367–85.

Jones, O. (1997), 'Little Figures, Big Shadows: Country Childhood Stories', in P. Cloke and J. Little (eds), *Contested Countryside Cultures: Otherness, Marginalisation and Rurality*, Routledge, London, pp. 158–79.

Kinsman, P. (1997), 'Re-negotiating the Boundaries of Race and Citizenship: the Black Environment Network and Environmental and Conservation Bodies', in P. Milbourne (ed.), *Revealing Rural Others: Representation, Power and Identity in the British Countryside*, Pinter, London, pp. 13–36.

Land Reform Policy Group (1999), *Recommendations for Action*, The Scottish Office, Edinburgh.

Lewis, G., McDermott, P. and Sherwood, K. (1991), 'The Counter-urbanization Process: Demographic Restructuring and Policy Response in Rural England', *Sociologia Ruralis*, vol. 31, pp. 309–20.

Little, J. (1998), 'Employment Marginality and Women's Self-identity', in P. Cloke and J. Little (eds), *Contested Countryside Cultures: Otherness, Marginalisation and Rurality*, Routledge, London, pp. 138–57.

Little, J. and Austin, P. (1996), 'Women and the Rural Idyll', *Journal of Rural Studies*, vol. 12, pp. 101–11.

Little, J., Clements, J. and Jones, O. (1998), 'Rural Challenge and the Changing Culture of Rural Regeneration Policy', in N. Oatley (ed.), *Cities, Economic Competition and Urban Policy*, Paul Chapman, London, pp. 127–45.

Lloyd, G. and Danson, M. (1998), 'The Land Question in Scotland – a Challenge for the New Parliament?', *Town and Country Planning*, vol. 67, pp. 364–5.

Lloyd, G. and Danson, M. (1999), 'A Land Reform Agenda in Scotland', *Town and Country Planning*, vol. 68, pp. 30–31.

Lowe, P., Ray, C., Ward, N., Wood, D. and Woodward, R. (1998), *Participation in Rural Development: A Review of European Experience*, European Foundation for the Improvement of Living and Working Conditions, Dublin.

Marsden, T. (1998), 'Economic Perspectives', in B. Ilbery (ed.), *The Geography of Rural Change*, Longman, Harlow, pp. 13–30.

Marsden, T. and Murdoch, J. (1998), 'Editorial: the Shifting Nature of Rural Governance and Community Participation', *Journal of Rural Studies*, vol. 14, pp. 1–4.

Marsden, T., Murdoch, J., Lowe, P., Munton, R. and Flynn, A. (1993), *Constructing the Countryside*, UCL Press, London.

McNicholas, K. and Woodward, R. (1999), *Community Development in North Yorkshire: An Assessment of the Objective 5b and LEADER II Programmes*, Research Report, Centre for Rural Economy, University of Newcastle.

Milbourne, P. (ed.) (1997), *Revealing Rural Others: Representation, Power and Identity in the British Countryside*, Pinter, London.

Morphet, J. (1998), *Rural Aspects of the Regional Agenda*, Briefing Paper, Local Government Association, London.

Moseley, M. (1984), 'The Revival of Rural Areas in Advanced Economies: a Review of Some Causes and Consequences', *Geoforum*, vol. 15, pp. 447–56.

Murdoch, J. (1997), 'The Shifting Territory of Government: Some Insights from the Rural White Paper', *Area*, vol. 29, pp. 109–18.

Murdoch, J. and Day, G. (1998), 'Middle Class Mobility, Rural Communities and the Politics of Exclusion', in P. Boyle and K. Halfacree (eds), *Migration into Rural Areas: Theories and Issues*, Wiley, Chichester, pp. 186–99.

Murdoch, J. and Marsden, T. (1994), *Reconstituting Rurality*, UCL Press, London.

Murdoch, J. and Pratt, A. (1993), 'Rural Studies: Modernism, Postmodernism and the "Post-Rural"', *Journal of Rural Studies*, vol. 9, pp. 411–27.

Newby, H. (1987), *Country Life*, Weidenfeld and Nicholson, London.

Pahl, R. (1965), *Urbs in Rure*, Weidenfeld and Nicholson, London.

Performance and Innovation Unit (1999), *Rural Economies*, Cabinet Office, London.

Philips, M. (1998a), 'Investigations of the British Rural Middle Classes: Part 1: From Legislation to Interpretation', *Journal of Rural Studies*, vol. 14, pp. 411–25.

Philips, M. (1998b), 'Investigations of the British Rural Middle Classes: Part 2: Fragmentation, Identity, Morality and Contestation', *Journal of Rural Studies*, vol. 14, pp. 427–43.

Philo, C. (1992), 'Neglected Rural Geographies: A Review', *Journal of Rural Studies*, vol. 8, pp. 193–207.

Pratt, A. (1996), 'Discourses of Rurality: Loose Talk or Social Struggle?', *Journal of Rural Studies*, vol. 12, pp. 69–78.

Ray, C. (1998), *New Places and Space for Rural Development in the European Union: An Analysis of the UK LEADER II Programme*, Working Paper 34, Centre for Rural Economy, University of Newcastle upon Tyne.

Ray, C. (1999a), *The Reflexive Practitioner and the Policy Process*, Working Paper 40, Centre for Rural Economy, University of Newcastle.

Ray, C. (1999b), *Reconsidering the Evaluation of Endogenous Development: Two Qualitative Approaches*, Working Paper 39, Centre for Rural Economy, University of Newcastle.

Rees, P., Stillwell, J., Convey, A. and Kupiszewski, M. (eds) (1996), *Population Migration in the European Union*, Wiley, Chichester.

Rural Development Commission (1999), *Rural Challenge: Lessons for the Future*, RDC, Salisbury.

Scottish Executive (1999), *Land Reform: Proposals for Legislation*, The Stationery Office, Edinburgh.

Seymour, S. and Short, C. (1994), 'Gender, Church and People in Rural Areas', *Area*, vol. 26, pp. 45–56.

Short, J. (1991), *Imagined Country*, Routledge, London.

Shucksmith, M. and Chapman, P. (1998), 'Rural Development and Social Exclusion', *Sociologia Ruralis*, vol. 38, pp. 225–42.

Sibley, D. (1997), 'Endangering the Sacred: Nomads, Youth Culture and the English Countryside', in P. Cloke and J. Little (eds), *Contested Countryside Cultures: Otherness, Marginalisation and Rurality*, Routledge, London, pp. 218–31.

Stillwell, J., Rees, P. and Duke-Williams, O. (1996), 'Migration Between NUTS Level 2 Regions in the United Kingdom, in P. Rees, J. Stillwell, A. Convey and M. Kupiszewski (eds), *Population Migration in the European Union*, Wiley, Chichester, pp. 275–307.

Tewdwr-Jones, M. (1998), 'Rural Government and Community Participation: The Planning Role of Community Councils', *Journal of Rural Studies*, vol. 14, pp. 51–62.

Townsend, A. (1991), 'New Forms of Employment in Rural Areas: A National Perspective', in T. Champion and C. Watkins (eds), *People in the Countryside. Studies of Social Change in Rural Britain*, Paul Chapman, London, pp. 84–95.

Townsend, A. (1993), 'The Urban-Rural Cycle in the Thatcher Growth Years', *Transactions of the Institute of British Geographers*, vol. 18, pp. 207–21.

Ward, N. and Lowe, P. (1998a), *Insecurities in Contemporary Country Life: Rural Communities and Social Change*, Working Paper 33, Centre for Rural Economy, University of Newcastle.

Ward, N. and Lowe, P. (1998b), *A 'Second Pillar' for the CAP? The European Rural Development Regulation and its Implications*, Working Paper 36, Centre for Rural Economy, University of Newcastle.

Ward, N. and McNicholas, K. (1997a), *Reconfiguring Rural Development in the UK: Objective 5b and the New Rural Governance*, Working Paper 24, Centre for Rural Economy, University of Newcastle upon Tyne.

Ward, N. and McNicholas, K. (1997b), *The European Union's Objective 5b Programmes and the UK*, Working Paper 28, Centre for Rural Economy, University of Newcastle upon Tyne.

Ward, N. and McNicholas, K. (1998), 'Reconfiguring Rural Development in the UK: Objective 5b and the New Rural Governance', *Journal of Rural Studies*, vol. 14, pp. 27–39.

Ward, N. and Woodward, R. (1998), *The Europeanisation of Rural Development Policy in the UK: The Case of East Anglia*, Working Paper 31, Centre for Rural Economy, University of Newcastle upon Tyne.

Weekley, I. (1988), 'Rural Depopulation and Counterurbanisation: A Paradox', *Area*, vol. 20, pp. 127-134.

Williams, G. and Bell, P. (1992), 'The "Exceptions" Initiative in Rural Housing – The Story so Far', *Town and Country Planning*, vol. 61, pp. 143–44.

Wood, D. (1998), *Globalization, Community Participation and Sustainable Rural Development: A Green Critique of EU Rural Development Policy*, CRE Working Paper 29, Centre for Rural Economy, University of Newcastle.

Woodward, R. (1996), 'Deprivation and "the Rural": an Investigation into Contradictory Discourses', *Journal of Rural Studies*, vol. 12, pp. 55–67.

Chapter Five

Leadership, Local Power and Rural Restructuring in Hungary

Imre Kovách

Introduction

This chapter is about issues of leadership, local power and rural restructuring in Hungary. It examines the socio-demographic and economic structure of rural Hungary and looks at the distinct meanings ascribed to the rural within a Hungarian context. Of particular interest are the economic, political and social processes established in the second half of the twentieth century and accelerated in the 1990s, which have moulded this part of Europe with reference to both global economic pressures and to internal moves to establish the necessary preconditions for Hungarian membership of the European Union (EU). Hungary is not an EU member state at present.

The transformation of the political regime after 1989 in Hungary prompted the reorganisation of most elements of its rural economy and society. Arable land was privatised (Harcsa et al., 1994, 1998a, 1998b; Kovács, 1994; Swain, 1994). Family farms, re-established cooperatives and new enterprises appeared to grow stronger and more robust. The previous socialist system for the local redistribution of resources and benefits was replaced by a welfare-orientated redistribution system more in line with the British welfare state model. Previously important forms of the agricultural second economy (private part-time farming, see Szelényi, 1988) such as the household plot system and state price guarantees, have disappeared (see Kovách, 1988, 1999a, 1999b; Szelényi, 1988). Rural society has restratified, new rural elite groups have emerged, those in the poorest social strata have become excluded, regional differences have increased and in 1996 a regional/rural development system was introduced to produce decentralisation and to conform to EU accession preconditions.

This chapter examines the process by which concepts of the rural and rurality have been reinterpreted in Hungary in the 1990s, at a time of dramatic changes in Hungarian rural society and economy. Images of the rural and rurality have shifted from their former positive sociocultural meaning to a situation where the rural and rurality have more problematic and negative connotations. In the past, rurality was considered a source of national culture; the rural was integral within national symbols and images, and the peasantry was viewed as a receptacle or custodian of national culture and the demographic basis of the nation. These images have been replaced by a more post-modern construction of the rural comprising a mixture of new and positive meanings, such as the rural as a place of recreation and the location of valued environmental attributes, and negative connotations, with the rural viewed as a problematic outcome of processes of modernisation and Europeanisation, and where rural localities are characterised as the home of the unskilled, the unemployed and the socially excluded. This chapter analyses the transformation of rural images as a process closely related to changes such as the disappearance of the historical peasantry, the decline of rural traditions, globalisation, shifts in systems for the redistribution of rural development resources, the changing function of the state and the emerging power of intellectuals and local elites. It examines these changes with reference to concepts of reflexive modernity and the culture economy as expounded by Bessiére (1998), Frouws (1998), Nygard and Storstadt (1998) and Ray (1998). I would argue that the analyses of these theorists can be extended with the concept of network analysis (Knorr-Cetina, 1981, 1994) and notions of class or group interests within a Hungarian context. This chapter will examine the networks of those actors who play a dominant role in the reinterpretation of the rural and rurality.

After the fall of communism, new actors emerged as new elites in local power networks within both local and national administrations and economies; these were representatives of and experts in the new regional/rural development system and comprised intellectuals who in effect have contributed to the construction of rural images. This shift from a monolithic communist system to a partially decentralised structure with multiple actors, and the introduction of a new regional/rural development system, are amongst the major changes in post-socialist Hungary. This chapter will examine the main features of these contradictory processes, which on the one hand have eliminated previously existing structures of local power and policy, but which on the other hand have prompted struggles and new challenges with the emergence of new hierarchies, weak democratic control and issues of participation.

The Socio-demographic and Economic Structure of Rural Hungary

The Definition of Rural and Demographic Change

In the second half of the twentieth century, the designation of a settlement as a town provided localities with advantages in the competition for state development resources under the socialist system for redistributive economic management. Because of this, local leaders would often use every means possible to acquire 'town' status for their settlements. Designating a settlement as a town was often the consequence of the political pressure exerted by interest groups, rather than any indicator of urbanisation as such. Many smaller towns retained their agricultural features in terms of their economic and occupational structure. It is probably closer to the truth to estimate the proportion of those living under rural living conditions in Hungary to be higher than the number and rate of villages or settlements which have 'village' administrative status. (See Table 5.1 for figures on population density in Hungary.)

Table 5.1   Territory, population and population density of rural Hungary, 1997

	Territory (km²)	Population	Population density persons per km²
Hungary	93,030	10,135,358	109
Rural regions	57,235	3,395,009	59
European Union*	3,230,800	372,000,000	115

Sources: KSH (Hungarian Central Statistical Office) datasets, 1997, *EUROSTAT datasets, 1997.

The population of villages had decreased consistently in the second half of the twentieth century until 1994, when this decline halted and the proportion of village population started to increase from 34.3 per cent in 1994 to 37.4 per cent in 1998 (see Table 5.2). New definitions adopted within Hungarian statistics in 1997 distinguished between basically rural, typically rural and urban regions. According to this categorisation, which applies European statistical (EUSTAT) methods, Hungary has a much more rural character than most EU member states with about 70 per cent of the Hungarian population living in basically or typically rural regions. Basically or typically rural regions are defined in this context from a total of 130 regions (under the first categorisation in 1997) or from 200-220 development regions (under the second categorisation after 1998) where the population of villages are dominant. It would be more accurate to state that about 50 per cent of Hungarians live in rural settlements, that is, farmsteads, villages, towns or cities with a strong rural character.

Table 5.2   The territory and population of rural regions in Hungary and the EU (1997)

	% basically rural	% typically rural	% urban	Total
Population				
EU member states	9.7	29.8	60.5	100
Hungary	33.5	40.1	26.4	100
Territory (%)				
EU member states	47.0	37.4	15.6	100
Hungary	61.5	34.6	3.9	100

Sources: European Commission, 1997; KSH (Hungarian Central Statistical Office) datasets, T-STAR database, 1997.

Hungarian settlement structure has not radically changed in the twentieth century. The proportion of the rural population relative to the total population decreased from about 60 per cent to 35–40 per cent by the late 1980s, but this decrease was generally understood as a consequence of the bureaucratic redesignation of settlements from villages to towns. The quality of life in the new towns and historic market towns and their social and economic structures hardly differed from that in the bigger villages, although (as described above) there has been a growth in population in rural localities. From the late 1960s to the 1990s, rural economic and social structure and the lifestyles in rural areas have been modernised as a result of investment in basic infrastructure, but there is still a characteristic gap between the living standards in urban and rural areas, with living standards lower in rural areas. As a consequence of the post-socialist transformation, Hungarian society has tended to be divided into urban and rural parts. As the figures in Tables 5.3 and 5.4 show, depopulation has threatened almost a third of all settlements, both urban and rural, and there are no clear differences in population change between urban and rural regions. Households move to rural areas from cities and rural families also search for better livings standards in towns. Counterurbanisation seems to be the strongest determining factor in population change, resulting in constant or increasing population levels in rural areas, but regional differences and the social and economic impacts of these are also causes of population change. Urban migration into rural areas commenced from the mid-1990s when economic liberalisation led to a growth in unemployment from 1990, to 12 per cent in 1994 and 1995, and to just under 10 per cent in 2000, and a deterioration in the living standards of the poorest urban strata and pensioners. The population of villages increased by 70,000 people (0.7 per cent of the total population) year by year after 1994. One of specific characteristics of counterurbanisation in Hungary is that the rural population has growth massively in urban areas (mainly in Budapest) while remote rural areas are endangered by out-migration. Counterurbanisation is a complex social change; the urban upper middle classes move to favoured leafy suburban settlements and those losing out in the post-socialist transformation hope to survive hard times in rural places where housing and living cost are lower than in cities. The North-East, some other Northern regions and small villages in micro-regions of Southern Transdanubia are mostly rural, where the population is ageing and some settlements are endangered by depopulation. A very problematic component of population processes is the growth of the very poor and excluded gypsy population and the emergence of gypsy villages in many rural regions.

Social and Economic Restructuring: Farmers, Entrepreneurs and the Unemployed

Economic Transformation

Privatisation, land restitution and economic liberalisation were key influences on rural restructuring in the 1990s. Of all sectors of the national economy, it was in agriculture that privatisation was most rapidly implemented. Rapid and radical agricultural privatisation had several causes, the most important of which was political considerations (rather than economic or social causes: see Csite and Kovách, 1997). The privatisation and transformation of agriculture, implemented in accordance with the political interests of the new elite, has replaced the entire political, social and economic environment of rural entrepreneurial activity.

Nearly two million families were entitled to restitution which meant that the amount of land available for restitution was insufficient to enable the

Table 5.3 Population changes 1994–97

Regions	Type of region	Population density, 1994, people/km²	Population density, 1997, people/km²	Natural reproduction 1994, per 1000 people	Natural reproduction 1997 per 1000 people	Ageing index 1994*	Ageing index 1997*	Net migration per 1000 people 1994	Net migration per 1000 people 1997
Budapest	Urban	3675.10	3544.41	-6.08	-6.70	148.56	160.58	-4.45	-6.63
Middle Hungary	Basically rural	85.36	85.84	-3.94	-4.69	104.00	107.41	4.98	6.36
	Typically rural	163.73	168.76	-1.69	-2.36	96.12	98.69	13.81	12.92
Middle Transdanubia	Urban	288.55	310.45	-0.48	-1.22	88.75	91.51	22.25	20.84
	Basically rural	56.81	56.69	-2.94	-2.76	93.60	99.36	-0.75	4.28
	Typically rural	111.72	111.36	-0.90	-2.29	89.70	99.80	2.23	1.30
West-Transdanubia	Urban	200.97	199.73	-1.41	-2.89	88.97	99.51	-0.22	2.28
	Basically rural	57.00	56.06	-4.53	-5.06	117.31	123.42	-4.35	0.06
	Typically rural	124.62	124.01	-2.57	-3.69	104.63	113.51	3.13	1.15
South Transdanubia	Basically rural	48.02	47.33	-4.13	-4.81	109.25	114.02	-2.23	-0.98
	Typically rural	107.27	105.68	-2.80	-3.76	106.43	115.46	1.42	-0.54
Northern Hungary	Basically rural	65.82	65.07	-3.07	-3.51	104.91	108.11	-3.60	0.32
	Typically rural	136.83	134.31	-2.57	-3.74	101.70	111.36	-2.34	-3.14
	Urban	274.99	270.12	-1.14	-2.42	90.47	99.94	-0.12	-4.7
Northern Plain	Basically rural	63.32	62.7	-1.28	-1.80	93.90	97.36	-5.78	-0.79
	Typically rural	159.6	159.11	0.04	-0.39	80.90	86.49	1.47	-2.42
Southern Plain	Basically rural	57.25	56.2	-5.16	-6.51	121.97	126.85	-2.42	0.57
	Typically rural	118.1	117.2	-2.21	-3.55	102.97	109.44	2.48	0.03

* Over 60, under 15%.

Sources: KSH (Hungarian Central Statistical Office); T-STAR database.

Table 5.4   Regional depopulation in Hungary (1997)

Regions	Number of depopulated settlements (1990s)	%
Middle Hungary	14	7.6
Middle Transdanubia	110	27.2
West Transdanubia	301	46.6
Southern Transdanubia	246	37.7
Northern Hungary	168	28.0
Northern Plain	69	17.8
Southern Plain	84	33.3
Hungary	992	31.7

Source: KSH (Hungarian Central Statistical Office); T-STAR database, 1997.

development of new, viable agricultural enterprises. Through restitution, 1.5 million households had become landowners by 1996. A significant proportion of rural society have become landowners and even many urban households have acquired land. However, more than 90 per cent of arable lands have been privatised, with a national average plot size for lands acquired by restitution of 4.4 hectares per household. It was out of the question that the ownership and production structures of agriculture before collectivisation would be restored by restitution (Kovács, 1994).

Private production is gradually becoming dominant in agriculture, although the number of registered individual entrepreneurs active in that sector did not grow after 1993. The number of registered individual farmers is about 30,000 (about 3-4 per cent of all family farms). There are about 1.2–1.6 million private family farms, the majority of which are part-time and mostly produce for household subsistence. The average area of land held by private family farms is below 1 hectare (Burgerné, 1996). In the 1990s the structure of land ownership has been characterised by holdings smaller than 5 hectares; 44.2 per cent of holdings are below this size (Burgerné, 1996). About half of the land cultivated by individual farms belongs to units of less than 10 hectares. 62.5 per cent of arable land is cultivated by tenants. In the case of farms of over 50 hectares, only 23–26 per cent are owned by farmers (Harcsa, 1995; Harcsa and Kovách, 1996). Hungarian agriculture consists of a mixture of farm types, including full- and part-time family farms, cooperatives and share- and limited liability companies. This complexity may prove to be permanent.

Shifts in patterns of ownership and of economic and political structures proceeded rapidly after 1989, while the organisation of family farm production, inherited from the period of collectivisation, was more resistant to complex changes. The pressure on farms to produce for subsistence has become stronger as a consequence of general economic depression, hindering the growth of specialised production (Nemes and Heilig, 1996). A significant aspect of the new family units are enterprises developed through necessity. The size of private agricultural units has not gone beyond family dimensions. The average number of people employed by the largest private farms does not exceed two or three people, and only 10–15 per cent of all the private family farms can be regarded as small or medium agricultural production units suited for commodity production.

Between 1992 and 1999 some differentiation emerged between the top family farms (see Harcsa, 1994, 1995; and Harcsa and Kovách, 1996 for details). With the concentration of land and production, the best family farms with significant capital were able to develop into real enterprises, even under the adverse economic conditions of the period 1992–99. Furthermore, the slow differentiation and concentration of agricultural private production has resulted in changes in the top producer's mentality and practices. Data from 1982 and 1995 stratification surveys (see Kovách, 1988) enable comparisons to be drawn between the types of units which have evolved. This indicates that the proportion of free market enterprises has grown from 9.6 per cent to 21.4 per cent and the proportion of peasant-type farms of mixed production structure decreased from 69.5 per cent to 52.6 per cent. This is evidence for the gradual expansion of the market economy, but the dominance of peasant-type units also suggests that, despite the concentration of production and the strengthening of entrepreneurial character, the complete transformation of the private sector in agriculture is still far from being accomplished. The system of credit and state subsidy has had little impact on boosting entrepreneurial activity in the face of the dominance of the peasant type of production (Harcsa and Kovách, 1996).

Rural services and industry as well agricultural enterprises exist in an economic and social space full of contradictions. In 1993, of the 659,749 non-agricultural enterprises actually in operation, 148,920 were registered in rural settlements. The colourfulness of the resources utilised (Kuczi, 1996), the often rather mixed product structure of the agricultural family units and the income based on pluriactivity of rural households suggests the renewal and continuity of the traditional structure of rural activities. The majority of agricultural family units have no specialised production and the entrepreneurs within services and industry are also unable to acquire adequate profits from only a single sector. The income from businesses or from shops is still often supplemented or replaced by the garden and land. The pluriactive household strategy remains dominant and stable and it is one of the surviving elements of traditional peasant economic activity.

Social Consequences

These structural changes in agriculture took place at the same time as a broader transformation crisis across Hungarian economy and society, and these factors, together with the reshaping of the formerly efficient local redistribution system (see Harcsa et al., 1994, 1998a, 1998b) has radically altered the living conditions of the inhabitants of rural settlements. In 1988 the number of persons employed by agricultural units was 1,028,000, a figure which fell to 326,000 by 1996, 31.8 per cent of the total in 1988. According to Andorka (1996), social differences changed rapidly in 1990s, and the differentiation and relative inequality between rural and urban populations progressed further. The household incomes of agricultural manual labourers and the self-employed are the lowest among the economically active (84 per cent of the average income), hardly higher than that of the unemployed (68 per cent) and less than that of people receiving an old-age pension (96 per cent). Rural unemployment is much higher than urban unemployment due to the reduction in agricultural employees and industrial unemployment hitting rural, commuting and unskilled labourers more than the average (Köllő, 1997). One of the consequences of rural unemployment is the spread of farming by obligation or economic necessity, and other economic activity and forms of behaviour on an unprecedented scale (see Laki, 1997; Simonyi, 1995; Tardos, 1992).

In the 1990s, the society of rural settlements was restructured with dramatic force and speed. One of the most conspicuous phenomena was the appearance of massive rural poverty and its new forms, identified by several researchers as the emergence of a rural underclass (Csite and Kovách, 1997; Ladányi and Szelényi, 1996; Laki, 1997). In 1993, masses of rural people simultaneously lost their jobs, the possibility of disposing their property and the institution of ancillary farming (Csite and Kovách, 1997; Harcsa et al., 1994). Experts on the emergence of poverty have predicted the appearance of belts of rural ghettos in northeastern and southwestern regions (Ladányi and Szelényi, 1997).

As a result of the restratification of rural society, the restructuring of the economy and the new spaces of local political authority, the possibilities for enterprises and of 'embourgeoisement' are contradictory. The rural underclass – many of whom are Gypsies – have fewer and fewer opportunities to legally supplement their incomes through paid employment and the receipt of benefits (Kertesi, 1997; Köllő, 1997). A quarter of the rural unemployed are now unable even to produce food at the lowest level for their own household (without even a hen at home) (Laki, 1997). The poor as well as several other groups within rural society seek to find ways to survive through the illegal economy (Laki, 1996).

The Europeanisation of Administrative Structures and the Decentralisation of Power

Administrative Structure of Hungary

Hungary is a relatively small European country. Its administrative structure has three levels: central government, 19 counties and Budapest, and 3,175 self-governing settlements. The political power of central government was weakened considerably following the post-socialist transformation and the introduction of a new regional and rural development system, but it plays a key role in the redistribution of development resources and in the control of the counties and self-governing settlements. More than two thirds of the yearly income of self-governing settlements come from governmental sources and EU development programmes managed by central government. The 19 county councils coordinate regional development programmes, tourism, social policy and public health systems. Budapest, a European metropolis, holds the same position and responsibilities as the counties. Some formal and informal benefits accrue to Budapest because of its total population (20 per cent of the Hungarian population) and its central location. (See Table 5.5 for population figures.)

The self-governing settlements are autonomous institutions according to the European Charter of Self Government. Hungarian legislation grants specific tasks to self-governing settlements, which can also define special functions. There are 223 cities, five of which are termed 'towns with county rights' with specific benefits and obligations. The 1,119 settlements without city status have populations of between 1,000–10,000. There are 704 villages with populations ranging from 500 to 1,000 and 1,019 small villages with fewer than 500 inhabitants.

In 1990, legislation was passed to enable 'change of administration' and all the settlements obtained the right to organise local government. Also in 1990, 3,100 settlements elected local governments which replaced the former 1,500 Socialist local councils. After this local election many settlement associations were organised with the purpose of pursuing regional/rural and economic development. From 1990 to 1995, the government and parliament

Table 5.5   The population and territory of the 19 counties and Budapest

County	Territory (km^2)	Population
Budapest	525	1,811,552
Pest	6,393	1,032,672
Fejér	4,359	423,531
Komárom-Esztergom	2,265	311,770
Veszprém	4,613	371,862
Gy_r-Moson-Sopron	4,089	424,507
Vas	3,336	266,411
Zala	3,784	293,233
Baranya	4,430	400,806
Somogy	6,036	330,261
Tolna	3,703	243,701
Borsod-Abaúj-Zemplén	7,247	729,965
Heves	3,637	322,629
Nógrád	2,544	216,538
Hajdú-Bihar	6,211	541,581
Jász-Nagykun-Szolnok	5,582	410,694
Szabolcs-Szatmár-Bereg	5,937	569,676
Bács-Kiskun	8,445	532,465
Békés	5,631	391,702
Csongrád	4,263	417,668
Total	93,030	10,043,224

Source: KSH (Hungarian Central Statistical Office); T-STAR database, 1997.

disbanded the Socialist system of centralised decision-making in regional/ rural development but a new system of regional/rural development policy had not been developed in its place. The government ran occasional (although sometimes very effective) development programmes. In this period, the government provided assistance for underdeveloped regions, initiated and developed special programmes for the northeastern regions (the most backward regions), established a Regional Development Fund, ring-fenced specific sums in the state budget for rural development, ran experimental programmes in order to build up a decentralised institutional structure for regional development, organised Regional and Territorial Councils of Rural Development, used funding under the PHARE regional development programme to extend the Integrated Regional Development Programme in Eastern Hungary, stimulated action projects between settlements and provided support for underdeveloped regions and settlements with high unemployment.

In 1992, a Regional Development Fund was established with the purpose of supporting backward regions and economic structural transformation. In 1993 an Act of Parliament defined a list of backward settlements and those settlements where the unemployment rate was 50 per cent higher than the county average and state programmes were developed to limit the consequences of economic restructuring. In this first phase of new regional policy, the government and its legislation provided significant resources for regional/rural development which, together with PHARE funding, were aimed at alleviating some of the problems of transformational crisis. However, the relative inefficiency of early development and support policy, combined with Hungarian aspirations for EU membership (and the 'Europeanisation' of structures and institutions associated with this) prompted demands for the implementation of a new, better and clearly organised regional/rural development system. After a combative political and social dispute, these demands were met by legislation passed in 1996. The new system modified some of the existing administrative structures.

The New Regional Development Policy (1996–?)

The new system. In 1996 the Hungarian parliament passed a Regional Development and Physical Planning Act, which transformed the decision-making process, decentralising it for the first time in Hungarian history. The decentralisation resulted in the emergence of multi-actor networks of regional/ rural development (RRD) (see Figure 5.3). In the structure of RRD, two new levels (micro and macro regions) were created to institutionalise cooperation between government, counties, local associations and local governments. The objectives of the Act were to assist the development of the market economy, to improve economic conditions and living standards, to strengthen the conditions for sustainable development, to reduce regional differences, to encourage initiatives by regional and local communities and to bring development policy in line with other national objectives. The new system, the first such system amongst the post-socialist states, adopted many existing elements of EU regional policy including decentralisation, the principle of partnership, subsidiarity, and application and programming procedures (Horváth, 1999). The new system was a source of new regionalisation and the creation of new development regions, because the decentralisation of resource distribution led to the formation of new institutional structures based on former administrative units (counties and self-governing settlements). A comparison between the maps in Figures 5.1 and 5.2 illustrates this.

© Kővári József, Márton Mátyás, Zentai László

Figure 5.1   Administrative map of Hungary showing counties

Figure 5.2   Administrative map of Hungary showing the Regional Development Councils

The institutional structure of the new regional policy consists of five levels: 3,175 local governments, 130 (later 220) micro-regions, 19 county regional development councils, seven regional development councils (at a macro level) and a National Regional Development Council, and the Ministry of Agriculture and Regional Development.

The activities of the county regional development councils require cooperation between the government and business interests. Members of these 19 councils include presidents of self-governing counties, mayors of the biggest cities, representatives of the Ministry of Regional Development and Environmental Policy (which after 1998 was known as the Ministry of Agriculture and Regional Development), representatives of the associations of local self-governing settlements and representatives of business associations. The seven regional development councils consist of the presidents of the county regional development councils, representatives of eight ministries, representatives of business associations and chambers of commerce and six representatives of self-governing settlements. The regional development councils establish agencies to take forward regional development programmes and handle Pilot Action Funds. Members of the national Regional Development Council are government ministers. The government has a specific fund to finance initiatives under the new regional development policy, including grants, interest-free credit and credit guarantee schemes, and mechanisms to allow financial participation in the regional development associations. The distribution of direct sources of RRD finance, in accordance with EU practice, is based on statistical data including GDP, and is allocated by application to the development councils. Other indirect sources of finance are allocated following application to individual ministries.

The Rural in the System

Initiatives and schemes for rural development. The definition of initiatives and schemes has been a lengthy process. Prior to 1996, underdeveloped settlements were targeted for receipt of regional development policy and central development resources, particularly for infrastructure investment. Many rural settlements were supplied with a telephone system, gas piping, sewerage and water supply but infrastructure development did not contain special elements of complex rural policy. The EU PHARE programme provided a subsidy of 800 million ECU for the period 1990–98 for regional/rural development (for projects such as agricultural modernisation, environmental improvements, micro-regional projects and assistance for small enterprises) but implementation of these projects was very often delayed. After 1996 and the introduction of the New Regional Policy, rural areas were considered an integral part of regional development and problematic regions (towns and neighbouring villages) were targets for regional development policy. Rural settlements could obtain development funds by acting in coalition within a micro-region, but funds were allocated on a competitive basis and under this system urban centres had more power, almost without exception. There were no separate resources for the purposes of complex rural development and three ministries (Ministry of Interior, Ministry of Agriculture and Ministry of Environment and Regional Development) competed for control of rural policy.

In 1998, a new government coalition came to power, and the Ministries of Agriculture and Environment and Regional Development were merged to become the Ministry of Agriculture and Regional Development. That same year, a new Department for Rural Development was established, which subsequently became the Department of Rural Development Projects. The new department's task was to produce development initiatives and to harmonise and control rural development. The agricultural and regional development lobbies together pushed forward a new phase in rural development policy. The new rural development department is actively preparing for EU harmonisation and EU accession, rather than developing a new rural development regime. The department is constantly in conflict both externally and internally within the Ministry, with interest groups and administrative units trying to have their complex rural development objectives and goals accepted.

In 1999 and 2000, the main rural initiative was preparatory programming for the EU SAPARD programme. Rural micro-regions were invited to set out their development programmes and about 90 per cent of them prepared acceptable proposals. This was the first chance for rural leaders (both administrative and business leaders) to formulate special, local development plans which have been harmonised with SAPARD initiatives. This activity has been the most advanced 'bottom-up' planning in post-socialist Europe. At the time of writing, because EU SAPARD resources have not been available until summer 2000, the only achievements to date have been preparatory work on the programmes to be undertaken, such as the formulation of local goals, the development of cooperative working methods, familiarisation with EU programmes and structures, and the writing of proposals.

The redistribution of direct and indirect regional/rural development funds is achieved via an application system which has contributed, to a great extent, to the rise of new actors in rural development. The administrative staff of the ministries involved, and experts involved in the process of developing

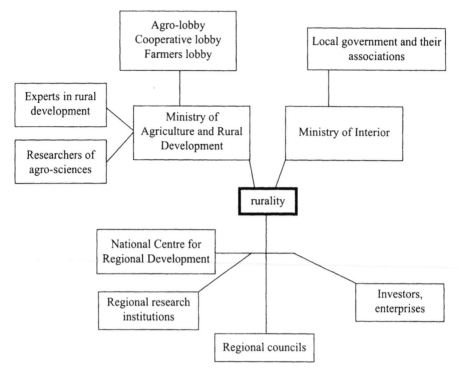

Figure 5.3   Actors in regional/rural development

Source: Csite and Kovách, 1999.

programmes and the application system, have control over the system and thus a monopoly on redistributive power.

The conflicts of the regional/rural development system. The new regional policy which was implemented in 1996 was a transforming system, which had to face many problems in terms rural development. There have been many problems. For a start, rural settlements are not powerful actors within the system. In the micro-regions, urban settlements play a central role and often exploit development resources. Small rural settlements are not able to compete in the application process, are unable to prepare proposals, to raise finance for match-funding or to lobby for funds in competition over PHARE and SAPARD finance. Rural development has been subordinated to regional development, despite the fact that 40 per cent of the population live in non-urban settlements. Furthermore, the system for the redistribution of resources for development is not well managed. One of the most important elements of

EU control and evaluation of its programmes, the monitoring of activities, is absent from the Hungarian system. Rural civil society is weak and does not have much control over the development system. Democratic control is also weak. Local economic elites have recognised that the regional/rural development system is virtually the only source of finance for economic development and in 1998 in parliamentary and local elections their representatives started attempts to assert political control over the process through legislation in local and regional development councils, in order to control the redistribution of funds.

In 1997, the government made a decision to implement programme monitoring, but at the time of writing (June 2000) it is apparent that this monitoring is still ineffective. The planned monitoring system does not conform to EU specifications, because in the Hungarian case, monitoring has been interpreted as a governmental task and state offices, ministries, and different councils within the regional system are proposed as the relevant authorities for monitoring. In other EU member states, autonomous institutions perform this work. Monitoring would control the redistribution of development resources in the absence of strong civil society and civil institutions. According to the plans for monitoring, the actors within regional/rural development will monitor each others' activities, and it is likely that this would in no way help to diminish existing rural disadvantages.

New Actors of Local Power: Participation, Autonomy, Local Elites and Civil Society

The Transformation of Local Policy

With the changes in the political system, which have secured broader political and human rights, people in rural areas can freely choose their national and local political representatives, their choice of jobs and their choice of place of residence. In respect of political power and local society, the omnipotence and primacy of politics is determined far less directly and the political alternatives for citizens and private individuals are much better, but the relationship between people and authority is not controlled by the community of private individuals and the public sphere.

In relationships between political authority and civil society, it is economic necessities which are decisive (Pálné, 1992). In Hungary, citizens can freely exercise their political rights, and there is no evidence of restriction of this.

Inequality between citizens in Hungary in the 1990s can be traced primarily to economic necessities caused by general impoverishment; it is impossible to view economic autonomy as separate from the autonomy of individuals, even in the case of bigger entrepreneurs. There are a number of reasons for the duality of formal political rights and the lack of real citizen's autonomy, including the aggressiveness of political authority in the face of civil society, which has been a feature of the twentieth century in various forms, and the authority of local governments which have emerged to replace previous systems of local redistribution, in which local elites have fought political battles with varying success. Because elected representatives in local government can be replaced in local elections, they can be controlled to a certain extent, but the relationship between the local authority and civil society in the form of private individuals does not change as a result (three-quarters of the mayors elected in 1994 were mayors in the previous round of elections as well) (Bocz, 1995, 1996).

Local Governments, Civil Sphere and Autonomy

Agrarian and rural policies have been of a rather aggressive nature particularly in the second half of the twentieth century; they left practically no space for autonomous civil institutions and the participation of local residents. After 1989, a general objective of the political parties has been to appropriate all kinds of civic initiatives and reorientate them on a party political basis. About 75 per cent of the budget of local governments derives from central sources, a proportion uniquely high in Europe (Berki and Orolin, 1996). The political parties of the post-socialist period are not particularly stable as political formations; voters' sympathies and choices of parties are volatile and unpredictable, which is quite irritating to the new administrative and entrepreneurial elite. It is perfectly understandable that the political elite has tried to subordinate civic organisations and movements to its authority by all means possible. The organisations of the civic sphere have been unable to avoid the parties and because of the crisis of transformation and its historical antecedents, civil society has been unable to support its own organisations. Therefore the sources of their finance are the foundations controlled by parties and by the elites of local government, and funds of the state are controlled by the state. In the first part of the 1990s about 20 per cent of the roughly 50,000 civic organisations operated in, or for the villages. About 80–85 per cent of civil organisations were founded by institutions, because the institutions of local society, the schools, hospitals and associations received a decreasing proportion of their funding from central sources. The foundations, associations and societies set up are primarily intended to mitigate the financial problems of institutions. The local organisation of budgetary orientation may effectively contribute to the development of some settlements, but these are by no means autonomous institutions. Their budget mostly comes from the state or local government. They may be effective institutions for the promotion of local culture, but through their operation it is the local elite which organises, legitimises and secures its authority and influence. Therefore these bodies are not promoters of local autonomy.

New Values and Diminishing Community Control

The linkages between economic and political influence on the part of the local elite have produced new hierarchies. It has been difficult to counter-balance these new hierarchies within civil society in the absence of appropriate civic institutions. The accepted values of rural communities, in which they were able to maintain communications between the individual and local representatives of authority (even in the socialist period) have dissipated as a result of both processes of globalisation and of systemic change. Although the end of state socialism has removed a series of bureaucratic and other obstacles from processes which ultimately will initiate a change in values, the lessons of the past 10 years for the countryside and rural society mostly concern the objectified set of values of consumer society between the pre- and post-modern period. Because there are no clear patterns, it is not possible to report any triumphant spread of individualisation, bourgeois mentality and associated values. The upper classes are inaccessible, far from the rural people, and there is nothing attractive in their world of values, beyond the influence of money and power. Material culture is the only sphere in which the middle strata of rural social groups, those occupying a relatively better position, have been able to keep pace with the world, even in the 1990s. This is the world of international consumer culture, flowing from television and other media carrying this message, but its force should be neither over- nor underestimated. This factor should be recognised; at present there is no other window open towards the world, and the values of the rural (and urban) population of more developed countries are increasingly determined by globalisation. In the Hungarian villages, the only obstacle to following Western material culture is money rather than any mental barrier, and with the progress of Western integration this openness and adaptability is becoming increasingly effective as a form of capital.

However, the 1990s is not only the period of the spread of consumer culture. The disintegration of traditional peasant communities, which started in the previous century, had not been completed when the introduction of the socialist system began. The values and patterns of behaviour of local societies have had rather strong community control until recently. This community control was always the means by which peasant traditions could be defended in the face of the alien and dangerous outside world. The force of these external influences can only be questioned by those who have not experienced the pressures to conform, where in a community of any self-respect one could not dress and consume, greet, celebrate and mourn just as one pleased until a few years ago. This was not an archaic, or ridiculously obsolete world and set of values, and its use of objects met all the expectations of its time at the end of the century. Its survival without the old economic background (with economic relationships based on private property) is likely to be due to the fact that rural people could still count on their ability to be protected by values regulated by community control. This was the only force keeping together local society in rural settlements during the period of rapid economic modernisation, when the functions of big families, the church, the school and other institutions of control and coordination disappeared. Despite there being many new initiatives for the reorganisation of rural identities, civil society and its institutions in the civil and public sphere are so disorganised that the rural renaissance is able to serve the purposes of local elite groups.

The Emerging Power of New Local Elites

A consequence of structural change is that the regional development and redistributive system and the key actors in economic development changed at the same time. In the political field of local and rural society, new social groups came into being, new elites which gained power and influence in the local governments, often for a long time. The new elite consists of an administrative elite (leaders of local governments) and the most powerful representatives of the economic elite. Local governments play a dominant role in the new redistributive and regional/rural development system, because of their role in the distribution of social and unemployment benefits, the control over the budgets of public institutions (schools, day nurseries, libraries, etc.) and local taxes. Although entrepreneurs are able to obtain an increasing proportion of development resources, local governments and the local/rural administrative elite still dominate over the use of resources (Oláh, 1996). This domination is important because 6–8 per cent of national income (280 billion Hungarian

Forinths), together with indirect sources totally 500 billion, is redistributed via the regional/rural development system. Local and regional elites are gaining increasing influence over the redistribution process for these resources. In the first half of the 1990s, infrastructure investment had priority and local governments and administrative elite groups were the most powerful actors in regional/rural development. Since 1996 (the first year of the New Regional/Rural Development System) employment projects received increasing amounts of money, and the chances of entrepreneurs (a local economic elite) have been augmented by the distribution of decentralised development resources. Credit and investment benefits can be obtained mainly from the regional/rural development system or from other state subsystems. These are primary sources of economic development because institutional actors in money-markets, banks, and savings schemes are not participating directly as investors in regional/rural development. One consequence of this is that regional/rural development money-markets, despite decentralisation, are policy-bounded and for this reason in the period 1997–99 economic elite groups started acquiring both finance and power in the local/rural political field directly or indirectly through their networks and exploiting the most important political positions. In 1998, in the campaign for parliamentary and local elections, this ambition led to the organisation of economic and political interests and lobbying groups in order to pick out and support candidates and to finance their electoral campaigns. The ambition of local economic and regional/rural elites to control political power and the representatives of local policy is particularly strong in small and medium-sized settlements when the possession of resources from the regional/rural development system cannot be accessed without the collaboration of local governments. The local elections in 1998 were influenced and decided by the articulation of the interests of local economic elites. The Hungarian political field doubled and local policy and the more prominent interest groups emerged, often in opposition to national policy. Local/rural autonomy is not strong and local interests have been structured within national policy networks. Within this particular structure of political relations the political networks, which are determined and utilised by a local/rural elite and which are able to control the regional development system through lobbying, are decisive tools for the exertion and influence of political power.

According to a comparative study (Csite and Kovách, 1999), actor networks of regional and rural development in Hungary basically differ from networks of actors in EU members states in which partnership control is dominant. The Hungarian system is in the process of developing and of evolving towards EU conformity and has arrived at a decisive phase which will determine whether a system similar to that of the EU will emerge, or whether a more diverse structure will come into being. The main problems for the Hungarian system are confusion over the purposes and strategy of regional/rural development, the subordination of rural development to regional development imperatives, and the hierarchy of actor networks. The Hungarian system has a decentralised decision-making processes but decentralisation has created double levels (nationwide and regional) of decision-making, while control and feedback mechanisms and institutions are missing. There are hierarchies in the network of actors which are dominated and controlled by decision-makers and the state apparatus. Those actors who are outside policy and state mechanisms are subordinated to political decision-makers. Civil control has low efficiency in the Hungarian regional/rural development system, which is confused and chaotic. In this construction of networks, local economic elites (top entrepreneurs in farming, industry and services) can influence the distribution of development funds in the following ways (see Csite and Kovách, 1999).

First, they can contribute to the formulation and reinterpretation of regional/ local identities and cultural values, which in turn can produce certain advantages in the competition for development funds. Economic elites can partner local administrative elites in the creation of non-cultural ideologies of employment policy and infrastructure investment. They can also maintain and strengthen partnership networks in order to influence local power. Second, they can influence decision-makers and experts (both through corruption and through networking) and become members of the political network hierarchy. Third, when this is insufficient, or where political linkages are hard to forge, the economic elite organises direct control of the decision-making process, by getting their representatives into the institutions of decision-making or by incorporating decision-makers within the hierarchical network of economic interest groups. Fourth, in the competition for development funds, the economic elites create hierarchical networks of clients so as to obtain profit via the redistribution of development funds and the disqualification of competitors.

If the system remains unaltered the economic elite will get power over the redistribution of development funds and over the decision-making process. Their political position and the consequences of this may transform hierarchical economic and political networks into basic structural elements which are potentially completely antagonistic with the objectives of democratisation and the movement to stabilise an unbalanced and therefore perilous political system and political process. According to field studies (Csite and Horváth, 1998) the local elites changed strategy and started political actions in 1998 at the time of local and parliamentary elections. A first step in the formation of a new interest group and in the manifestation of new political intentions was the development of a new discursive strategy. Representatives of the economic elite exerted much influence over local newsletters, TV and radio channels and in other similar public forums. They looked for political alternatives to the socialist-liberal government parties, which had not facilitated their access to privatised property and development resources, and in this political struggle they identified their political and economic interest group as representative of the local interests of the entire local/rural community. The kinds of reforms which were implemented have been analysed with reference to the language of local newsletters, which tended towards localism. The administrative elite, which is simultaneously threatened by and partially preparing for EU accession, has been searching for development resources and possible arguments and motives of successful applications. Many local history books have been published, local notables have been rediscovered, local monuments have been dusted and polished, (new) old local dishes have been cooked, wines have been bottled and the post-socialist rural world seems to be tending towards a state of reflexive modernity, in which image making, discursive strategy and the power of intellectualism have much more importance.

Competing Rural Images

In this chapter, I adopt Ray's thesis on intellectual property and the emerging power of intellectuals in rural development (Ray, 1998). In Hungary, 1998 saw the emergence of a discourse concerning the scientific and political interpretation of rural, which referred mainly to the necessity of introducing measures to conform to EU regional/rural development policy. The discussion about the definition of rural is more political than scientific or intellectual. The discussants' aim is to influence the redistribution of development resources and their arguments represent different political interests and other lobbies (Csite and Kovách, 2000). As a consequence of the Hungarian political tradition and the present condition of political and intellectual life, rural images with a figurative sense are much more legitimate and express different intentions of the various actors in rural development who both direct and dispute initiatives and the redistribution of rural development resources. Intellectuals and experts who design and disseminate rural images dispose effective power over rural development. They are new and influential actors in rural development.

As mentioned briefly above, the interpretation of the rural (and of the urban) was related to competition for development resources in the communist period. With changes in political structures in the 1990s, the interpretation of 'rural' and of rural images is increasingly subordinated to development policies and the development system. Under the communist regime, success in competition for development funds was part of an indistinct lobbying process, and the exposition or manifestation of rural or local/regional images was not a prior condition of development. Although positive rural images have never had enormous influence, amongst socialist technocrats, negative rural images effectively influenced the decision-making processes. In the 1970s, for example, a group of urbanists (political and academic representatives of urban pressure groups) asserted that small villages were not viable. A consequence of this negative rural image was the priority granted to regional cities and some small depopulating village regions in development policy.

Rural Images: Joining National and Rural and Criticism of the Rural Idyll

In the long history of rural images and image design, from nineteenth century romanticism to contemporary post-modernity, a dominant fact in the production of rural imagery has been its policy orientation. The interests of intellectuals in folk art and rural life was spurred by the predictions of Herder, the theorist of the German Enlightenment, on the ruin of nations without folk art in the nineteenth century. This intellectual movement led to the formation of many rural images in the last century which had a distinctive impact on Hungarian history. Surviving romanticism, they have been incorporated into modern and postmodern rural images. From the variety of factors influencing nineteenth century rural images, two components emerged as dominant in the twentieth century, namely the articulation of relationships between national and rural imagery, and criticisms of the notion of the rural idyll, which originated in the Reform Era of first half of 1990s when noble reform politicians rushed to supported the liberation of serfs (peasants within the feudal system).

Following the intellectual and political movements of the nineteenth century and the search for national identity, the rural was identified as a social space for the nation where gentry, landlords and peasants together were represented as embodying the spirit of the Hungarian people, in comparison with the ethnically mixed urban areas (with their German, Jewish, Slavic and Hungarian populations) and the international aristocracy. From the middle of the nineteenth century, so called 'Folkism' was the dominant ideology and trend in literature and art, producing critical but basically positive rural images ranging from representations of the *puszta* (plain), the *shadoof* (a system for raising water deploying a counterweighted bucket on a pole), the vineyard and the Balaton lake landscape, through peasant literary genres and the role of peasant heroes, to the decorative motifs of Art Noveau. After the fall of the gentry and the traumas of the First World War, peasant culture and the peasantry (both taken as rural features) were taken as a base of the nation and national culture. 'Just from pure origin', a motto used by the composers and musicologists Béla Bartók and Zoltán Kodály, was taken up by generations of intellectuals who were committed to social and cultural reforms.

The national myths of rurality were associated with criticism of conditions of rural and peasant life from the early nineteenth century, but this intellectual criticism and occasional political action for peasants and rural areas became stronger after the defeats of the First World War when the voices of nationalists emerged, coupled with those of a new generation of intellectuals with origins in peasant families in the countryside, who interpreted their historical mission as the radical reformulation of rural economy and society, Hungarian culture and its social structure. This generation, later called the 'Folk-writers movement', made its first appearance with impressive sociological, ethnographical works, literary reports and sociographies in which they replaced notions of the rural idyll with images of poor peasantry, rural misery, rural social conflicts, economic and health problems, and claims for land distribution.

The Decline of the Peasantry, Communists and Rural Images

The traditional peasantry which had formed a dominant social strata, demographically, within Hungarian society in the middle of the twentieth century has transformed into different social entity in the second half of that century. The decline of the peasantry in Hungary was a consequence of repeated attempts to adapt and modernise the rural economy, something which has resulted in state and policy interventions in the rural economy and society. These include the protection of big estates in the first decades of the twentieth century, the introduction of a war economy and increasing state control during the time of World War II, the three radical reforms for restructuring land redistribution from 1945 (two phases of collectivisation from 1948-1958 and 1958–63 and the Kadarist modernisation of the 1960s and 1970s), and reprivatisation and land restitution after 1989 (Kovách, 1999a, 1999b, 1999c).

Policy intervention into rural structures and rural life was extremely strong during Communist rule (see Harcsa et al., 1998a, 1998b; Kovách, 1999a; Meurs, 1999) when an adopted Soviet model gave preference to industry and the working class, enforced collectivisation, industrialisation and urbanisation, and which held dominance over cultural life and the construction of rural images. There is no space in this chapter to analyse changing rural images in the communist period, although it should be noted that this was a complex and contradictory process of diverse periods with the medium of such rural images ranging from art to film, via propaganda posters. On the one hand, rural images served the political interests of communist politicians, for example, to agitate for collectivisation and against *kulaks* (rich peasants), who created (or at least tried to create) new legends and heroes of the communist epoch. On the other hand, during the period the role of the peasantry in history was reinterpreted, and in the 1950s very popular historical films were made, focusing on peasant characters. In the late 1950s and 1960s, the best works of film and literature presented a realistic overall view of rural transformation showing the distress caused by collectivisation and the decline of the peasantry. From the late 1960s and early 1970s, an increased critical focus in film and in the fine arts was used in the portrayal of rural life and society, and scholarship in the social sciences actively contributed to the framing of rural images. From 1950s to the fall of communism, the dominant rural images were more positive than negative, although criticism, realism and documentarism were the determinative style of cultural products. The communist regime utilised some rural images from the pre-war period (for example, the agitating force of nationalism, and the articulation of relationships connecting notions of nationhood to rurality). Their political opponents also showed a preference for some elements of traditional rurality and peasant culture. As a consequence, historic elements within rural images survived communism, despite changes in rural realities and the disappearance of the traditional peasantry. I would also emphasise that the making of rural images was not localised; at a time of centralised political power, locality had no representation in policy and culture.

Rural Image(s) in Post-socialist Times

Rural images have undergone radical changes in post-socialist times. While cinema, fine arts and social sciences from the 1950s to the late 1980s contributed effectively to rural image-making, this role has been replaced in the 1990s by mass media with the production of more realistic images of the rural emphasising negative and problematic aspects. Modern mass media and news magazines concentrate on extravagant stories, free from insightful comment, in which the rural has negative meanings because of the focus in such stories on scandals such as the problems of local government, difficulties in schools, the problems faced by the gypsy minority, the reprivatisation of land and so on. The consequence is that the rural in mass media and in public opinion is taken to mean backward, underdeveloped, poor, disgraced and repulsive.

The change of regime caused a vacuum in the media, later filled with elements of postmodern ideology. Celebrities, speakers and mass media reporters adopted postmodern media discourses, manners and feelings. The postmodern has no utilisable elements of imagery for and about the post-socialist rural. Another consequence of the competition to fill the post-socialist media vacuum is that elements of existing rural images have been incorporated into postmodern image panels, in turn leading to creative chaos where the various components of dominant rural images have lost their original meanings. In the cavalcade of post-modern mass media, rural images are diverse and indicate the lack of intellectuals familiar with concepts of postmodernism, who might otherwise be capable of creating competing rural images.

Working on a textbook of rural images in collaboration with my research partners and students (Kovách and Saád, 2000), we have identified the following rural images in policy, media, literature.

1   There are political images which serve political interests; this was particularly important at the time of the privatisation and market re-orientation of agriculture.
2   The language of smallholders often appears in the terminology of leaders of the second government party (from which the Minister of Agriculture and Rural Development is selected). This language and discourse combines elements of folk plays and operettas with light populism and nationalism.
3   There are also images of rural space as the home of simple, resilient folk, represented as inferior to urban dwellers.
4   Anti-urban feelings and sentiments are often found.
5   The rural can often be portrayed as an archaic society, a notion evident in the discursive strategy of the Church and the folk intelligentsia.
6   The rural may also be seen as a space for the improvements and activities of 'experts'.
7   The rural may be idyllic and bucolic.
8   There is also the rural as an area of tourism, often associated with specific culinary traditions.
9   The rural and the agricultural are often connected.
10 Nature can be seen as a source of recreation.
11 The use of caricatures has been the primary approach of liberals to rural minorities, particularly the poor, women and older people.

All these rural images gained new importance after 1996 when a new system of regional/rural development was implemented and after the competition for development resources prompted local elites to search for ways of localising the use of development resources and putting into action new discursive strategies for the creation and elaboration of local/rural images. Early indications suggest that this might be successful, along the lines described by Ray (1998), Bessiére (1998), Nygard and Storstadt (1998) and Frouws (1998) with reference to rural development in other EU member states. The 'intellectualisation' of rural development, which ultimately may lead to reflexive modernity in the post-socialist rural and amongst rural people, will strengthen the class position of designers and intellectuals who are powerful actors in rural development.

Conclusion

The different questions and problems that have been raised in this chapter show that leadership, local power and post-socialist rural restructuring in Hungary is far from being balanced or completed in spite of the fact that decentralisation in the redistribution of development resources has been partially achieved, and the institutions of a market economy have been introduced. This story may speak about the play of new actors in surviving structures of power. In the processes of democratisation and Europeanisation, new actors and their networks have appeared in local power configurations, and the former monolithic power structure has been replaced by multi-actor networks. Centralised political power has lost its former omnipotence, but is still evident from the collapse of communism to civil control over politics. Powerful groups of administrative and economic elites have emerged to take positions in the redistributive system, the system which still has most influence on rural development in spite of full institutional transformation. The leadership and power relations of rural development have started to be more intellectual with the evolution of politically impressive rural images. The need to design rural development and application systems has offered greater potential than before for intellectuals and experts to represent their financial, political and class interests. Elsewhere I have argued that EU accession will contribute to profound changes in the present status of Hungarian (rural) policy (Kovách, 2000). On the other hand, it is also the case that EU business can contribute towards the emergence of new hierarchies of power.

References

Andorka, R. (1996), 'A Társadalmi Jelzőszámok Tükrében' ('In the Mirror of Social Indices'), *Társadalmi riport, 1996*, Tárki, Budapest, pp. 16-43.

Berki, E. and Orolin, Z. (1996), 'Az Önkormányzati Szövetségek Helye az Érdekegyeztetésben' ('The Place of Associations of Local Governments in Interest Adjustment'), *Társadalmi Szemle*, vol. 51, pp. 79–88.

Bessiére, J. (1998), 'Local Development and Heritage. Traditional Food and Cuisine as Tourist Attraction in Rural Areas', *Sociologia Ruralis*, vol. 38, pp. 22–35.

Bocz, J. (1995), 'Önkormányzati Képviselők, Polgármesterek' ('Representatives of Local Governments and Mayors'), *Társadalom-statisztikai füzetek*, 13, KSH, Budapest.

Bocz, J. (1996), 'Azönkormányzatok Döntéshozói' ('Decision-makers of Local Governments'), *Társadalom-statisztikai füzetek*, 13, KSH, Budapest.

Burgerné, G.A. (1996), 'A Magyarországi Földpiac' ('The Land Market of Hungary'), *Statisztikai Szemle*, vol. 74 (5–6) (May–June), pp. 411–20.

Csite, A. and Horváth, G. (1998), *A Vállalkozók a Helyi Társadalomban* (*Entrepreneurs in Local Society*), unpublished manuscript.

Csite, A. and Kovách, I. (1997), 'A Falusi Szegénység' ('Rural Poverty'), *A falu*, vol. 1.

Csite, A. and Kovách, I. (1999), *A Regionális és Vidékfejlesztési Politka Európaizálása Magyarországon az 1990 – es években* (*Europeanisation of Regional and Rural Development Policy in Hungary in the 1990s*), unpublished manuscript.

Csite, A. and Kovách, I. (2000), *A Falusi Társadalom Átalakulása* (*The Transformation of Rural Society*), unpublished manuscript.

European Commission (1997), *CAP 2000 Rural Developments*, Brussels.

Frouws, J. (1998), 'The Contested Redefinition of the Countryside. An Analysis of Rural Discourses in The Netherlands', *Sociologia Ruralis*, vol. 38, pp. 55–69.

Harcsa, I. (1994), *Parasztgazdaságok, Mezőgazdasági Vállalkozók* (*Peasant Farms, Agricultural Entrepreneurs*), KSH, Budapest.

Harcsa, I. (1995), *Farmerek, Mezőgazdasági Vállalkozók* (*Farmers, Agricultural Entrepreneurs*), KSH, Budapest.

Harcsa, I. and Kovách, I. (1996), 'Farmerek és Mezőgazdasági Vállalkozók' ('Farmers and Agricultural Entrepreneurs'), *Társadalmi Riport*, 1996, Tárki, Budapest.

Harcsa, I., Kovách, I. and Szelényi, I. (1994), 'A Posztszocialista Átalakulási Válság a Mezőgazdaságban és a Falusi Társadalomban' ('Postsocialist Crisis of Transformation in Agriculture and in Rural Society'), *Szociológiai Szemle*, vol. 3, pp. 17–31.

Harcsa, I., Kovách, I. and Szelényi, I. (1998a), 'The Hungarian Agricultural Miracle and the Limits of Socialist Reforms', in I. Szelényi (ed.), *Privatising the Land: Rural Political Economy in Post-Socialist Societies*, Routledge, London, pp. 21–43.

Harcsa, I., Kovách, I. and Szelényi, I. (1998b), 'The Crisis of Post-Communist Transformation in the Hungarian Countryside and Agriculture', in I. Szelényi (ed.), *Privatising the Land: Rural Political Economy in Post-Socialist Societies*, Routledge, London, pp. 214–45.

Horváth, G. (1999), 'European Access and Changing Hungarian Regional Policy', in A. Duró, (ed.), *Spatial Research in Support of European Integration*, Centre for Regional Studies, Kecskemét, pp. 13–31.

Kertesi, G. (1997), 'A Gazdasági Ösztönzők Hatása a Népesség Földrajzi Mobilitására 1990 és 1994 Között (Település- és Körzetszintő Elemzés)' ('The Influence of Economic Incentives on the Geographic Mobility of the Population Between 1990 and 1994. An analysis at the Level of Settlements and Districts), *Esély*, vol. 2, pp. 3–32.

Knorr-Cetina, K. (1981), *The Manufacture of Knowledge*, Pergamon Press, Oxford.

Knorr-Cetina, K. (1994), 'Primitive Classification and Postmodernity: Towards a Sociological Notion of Fiction', *Theory, Culture and Society*, vol. 11, pp. 1–22.

Kovách, I. (1988), *Termelők és Vállalkozók. A Mezőgazdasági Kistermelők a Magyar Társadalomban* (*Producers and Entrepreneurs. Small-scale Agricultural Producers in the Hungarian Economy*), Társadalomtudományi Intézet, Budapest.

Kovách, I. (1999a), 'The Lessons of History for Future Rural Development: Agriculture, Rural Society and State Policy in 20th Century Hungary', in P. Starosta, I. Kovách and K. Gorlach (eds), *Rural Societies Under Communism and Beyond: Hungarian and Polish Perspectives*, Lodz University Press, Lodz, pp. 17–44.

Kovách, I. (1999b), 'The Post-socialist Hungarian Countryside: Transformations in 1990s and the Alternatives', in P. Starosta, I. Kovách and K. Gorlach (eds), *Rural Societies Under Communism and Beyond: Hungarian and Polish Perspectives*, Lodz University Press, Lodz, pp. 361–72.

Kovách, I. (1999c) 'Hungary: Co-operative Farms and Household Plots', in M. Meurs (ed.), *Many Shades in Red: State Policy and Collective Agriculture*, Rowman and Littlefield, London and Boulder, pp. 126–51.

Kovách, I. (2000) 'LEADER, a New Social Order, and the Central and East European Countries', *Sociologia Ruralis*, vol. 40, pp. 181–90.

Kovách, I. and Saád, J. (eds) (forthcoming) *Rural Images*.

Kovács, T. (1994), 'Térségi Sajátosságok a Földkárpótlásnál' ('Regional Specificities in Land Restitution'), *Agrártörténeti Szemle*, vol. XXXVI, 1–4, pp. 77–87.

Köllő, J. (1997), 'A Napi Ingázás Feltételei és a Helyi Munkanélküliség Magyarországon. Számítások és Példák' ('Conditions of Daily Commuting and Local Unemployment in Hungary. Calculations and Examples'), *Esély*, vol. 2, pp. 33–61.

Kuczi, T. (1996), 'A Vállalkozók Társadalmi Tőkéi az Átalakulásban' ('Social Capital of Entrepreneurs in the Transformation'), *Századvég*, summer.

Ladányi, J. and Szelényi, I. (1996), *Making of a Rural Underclass; the Gypsy Village Ghetto and the Poor of Csenyéte*, unpublished manuscript.

Ladányi, J. and Szelényi, I. (1997), 'A Társadalom Etnikai, Osztály- és Térszerkezetének Összefüggései az Ezredforduló Budapestjén' ('Relationships of the Ethnic, Class and Spatial Structure of Budapest Society at the Turn of the Millennium'), in Z. Kárpáti (ed.), *Társadalmi és Területi Folyamatok az 1990-es évek Magyarországán*, MTA TKKK, Budapest.

Laki, L. (1996), 'Töredékes Helyzetkép a Perifériáról' ('A Fragmentary Image of the Situation at the Periphery'), *Társadalmi Szemle*, vol. 51, pp. 3–21.

Laki, L. (1997), 'A Magyar Fejlődés Sajátszerőségeinek Néhány Vonása Avagy a Polgárosodásból Kimaradó Társadalmi Csoportok' ('A Few Specific Features of Hungarian Development/or the Social Groups left out of Embourgeoisement'), *Szociológiai Szemle*, No. 3, pp. 67–91.

Meurs, M. (ed.) (1999), *Many Shades in Red: State Policy and Collective Agriculture*, Rowman and Littlefield, London and Boulder.

Nemes, G. and Heilig, B. (1996), 'ÖneLlátás és árutermelés. Mezőgazdasági Kistermelők egy Észak-magyarországi Faluban' ('Subsistence and commodity production. Small-scale agricultural producers in a North Hungarian village'), *Szociológiai Szemle*, 3–4, pp. 149–80.

Nygard, B. and Storstadt, O. (1998), 'Deglobalization of Food Markets? Consumer Perception of Safe Food: the Case of Norway', *Sociologia Ruralis*, vol. 38, pp. 36-54.

Oláh, M. (1996), 'A Helyi Elit Szerepe a Kelet-közép-európai új Demokráciák Önkormányzati Életében' ('The Role of the Local Elite in the Life of Local Governments of the East Central European New Democracies'), in Z. Agg (ed.), *Átépítés, Közigazgatás – Területfejlesztés – Városmarketing*, Veszprém Megyei Önkormányzat, pp. 215–45.

Pálné Kovács, I. (1992), 'A Helyi Hatalom Határai' ('The Limits of Local Government'), in F. Csefkó (ed.), *Helyi Társadalom, Gazdaság, Politika. Tanulmányok az Önkormányzatokról*, Alapítvány a Magyarországi Önkormányzatokért, Budapest, pp. 60–72.

Pálné Kovács, I. (1993), 'A Lokális Anatómiája' ('The Anatomy of the Local'), in I. Pálné Kovács and F. Csefkó (eds), *Tények és Vélemények a Helyi Önkormányzatokról*, MTA RKK, Pécs, pp. 7–33.

Ray, C. (1998), 'Culture, Intellectual Property and Territorial Rural Development', *Sociologia Ruralis*, vol. 38, pp. 3–20.

Simonyi, Á. (1995), 'Munka Nélkül' ('Without Work'), *Szociológiai Szemle*, vol. 1, pp. 55–70.

Swain, N. (1994) 'Transition from Collective to Family Farming in the Post-Communist Central-Eastern European Countryside', *Eastern European Countryside*, pp. 17–30.

Szelényi, I. (1988), *Socialist Entrepreneurs*, University of Wisconsin Press, Madison.

Tardos, K. (1992), 'Marginális Csoportok a Munkaerőpiacon: a Munkanélküli Segélyből Kizártak' ('Marginal Groups in the Labour Market: Those Excluded from Unemployment Allowance'), *Szociológiai Szemle*, vol. 1, pp. 99–109.

Chapter Six

Rural Restructuring, Power Distribution and Leadership at National, Regional and Local Levels: the Case of France

Nicole Mathieu and Philippe Gajewski

Introduction

France is well known for two characteristics: an enduring self representation as a 'peasant society',[1] even though farmers had become a minority in national and rural society before the last war; and the strength of a micro-local tradition of territorial administration originating in the Revolution (the 36,000 communes), which resists even if decentralisation from the beginning of the 1980s has increased 'intercommunalité'. These peculiar features are highly influential politically and help to shape the character of rural restructuring in France. A decrease in agricultural power has generated a reaction to maintain the realisation at the national and EU level of the dangers of rural 'desertification'.[2] In addition, a rural renewal by positive net migration and the 'ruralisation of industry' has seen the development of new local leaders tending to counterbalance urban political power. Rural and urban representations and controversies have played a considerable role in the resulting redistribution of power in rural France, as they have more specifically in French and EU development policies.

This chapter starts by providing some basic contextual information on rural France, considering demographic changes, economic developments and employment trends, and patterns of urbanisation and counterurbanisation. Consolidating this background overview, the second section provides a brief description of the principal administrative structures governing the management of rural areas at national, regional and local levels. The third section then gives a description of recent EU and national policy initiatives for rural economic and social development. The focus here is to understand the articulation of these programmes, such as Objective 5b funding and LEADER, with French 'local development', which has sought since the 1960s to create new local policies and to incorporate elements of popular or local participation within wider existing policies and administrative frameworks. The fourth section looks at the influence of debates over the meaning of 'ruralité' within policy discourses on development, as well as the influence of emerging policy networks and new social movements on development policy in both the social and agricultural spheres. The section examines rural restructuring, power and leadership at both regional and local levels, looking at the roles of different actors at the local level in the rural development process. The conclusion explores the possibility of theorising the relation between local/rural development policies and empowerment.

Rural Socio-demographic and Economic Processes in France, 1950-2000: From Rural Exodus to Rural 'Vitality'

Problems of Definition

The statistical definition of rural units is highly controversial in France. The official view of INSEE[3] has moved away from the traditional definition of a rural commune (fewer than 2,000 inhabitants agglomerated in the centre). INSEE wanted to prove the irreversible extension of urban areas, in terms of both the increasing density of buildings and the development of mobility and commuting. With rural areas being characterised by agricultural workers and primary production, new employment and industrial workers could not be 'rural', and rural areas were restricted to those of demographic decline and increasingly aged populations. At every census, rural units (using the traditional definition) were declared urban units or ZPIU (Zones de Peuplement Industriel et Urbain).[4] This tenacious INSEE representation led to a statistical division. On the one side was a positive type of space continuously in territorial expansion, characterised by polarity, urban poles were heads of developed networks creating peri-urban and multi-polarised areas (espace à dominante urbaine). On the other side was a negative type of space, defined as 'not under urban influence'; these had a lower degree of polarisation and, as a consequence, were isolated and statistically in regression (espace à dominante rurale).

Table 6.1 presents the populations of the latest INSEE partitions of French space. Many divisions, such as 'peri-urban communes', 'rural under weak urban influence', 'periphery of rural poles' and 'dominant rural area', have no administrative meaning and are defined solely by population characteristics. That is why we call them 'type of space'. They are a sort of official

Table 6.1   Settlement and employment indicators by types of space, 1990

Size of (influential) urban pole		Population density (people/km²)		Urbanisation rate (% population in urban units)		Employment rate (% employment/ active population)
		1990	Change 1982–90	1990	Change 1982–90	1990
Urban poles	5,000–19,999	305.8	+1.9	100	0.0	121.8
	20,000–99,999	755.3	+1.5	100	0.0	121.4
	100,000+	1784.0	+4.2	100	0.0	110.9
Peri-urban communes	5,000–19,999	50.9	+10.1	12.5	-3.2	48.3
	20,000–99,999	73.1	+12.7	26.9	-2.0	50.6
	100,000+	94.5	+19.7	49.3	+0.3	59.1
Dominant urban area		*269.5*	*+5.3*	*86.3*	*-1.3*	*102.5*
Rural under weak urban influence	5,000–19,999	36.2	+3.1	18.0	+0.2	73.3
	20,000–99,999	44.4	+4.4	31.8	+0.8	75.5
	100,000+	36.1	+7.4	31.8	+0.9	73.0
Rural poles		154.3	-0.9	99.3	-0.1	130.9
Periphery of rural poles		29.7	+4.6	5.2	-2.8	56.5
Remote rural		25.0	-2.5	21.7	+1.4	93.3
Dominant rural area		*34.9*	*+0.9*	*34.3*	*-0.1*	*88.9*
Total France		**104.1**	**+4.2**	**74.0**	**-0.6**	**99.5**

Source: INRA and INSEE, 1998, p. 31.

representation of the main divisions of French territory. Nevertheless, they do have an objective influence upon the administrative context and rural policies because French national and regional planning has always promoted a territorial way of acting. For instance, 'aménagement du territoire' (area development) policy differs from 'regional planning': in the former there should be a congruence between territorial management and 'type of space' management. The INSEE (negative) definition of rurality has played a role[5] in rural development policy since the 1960s (Bontron et al., 1983), positioning it in opposition to other types of spatial policy, such as metropolitan and town policies.

The Rural Demographic Revival Beginning in the 1970s

In spite of INSEE's statistical reduction of the rural territory,[6] a reverse trend could be observed in French rural areas beginning in the mid-1970s. The secular trend of 'rural exodus' stopped and even remote rural areas registered net in-migration. This reverse was detected quite early by Bontron and Mathieu (1968, 1973; Mathieu, 1974), who proposed the concept of 'rural space' to draw attention and designate the new processes and functions emerging. Affecting low density and mountainous areas as well as peri-urban areas, the phenomenon was not called counterurbanisation, as in Britain and the USA, but 'rural repopulation' (Bontron, 1983–84, 1989), 'rural renaissance' (Kayser, 1990), 'post industrial ruralité' (Jollivet, 1997) or 'new ruralité' (Mathieu and Robert, 1998), depending on the main phenomenon considered the cause of this reverse. At the same time, the first manifestation of a depopulation of city centres was called an 'urban exodus'.

As INSEE official presentations always tend to minimise the rural revival as metropolitan centres decline, it is quite difficult to give a precise comparative idea of this evolution by type of space, distinguishing inner Paris and outer Paris, inner principal metropolitan cities as Lyon and Marseille and their outer areas, and other metropolitan districts from different types of rural units (cf. the SEGESA[7] data base and publications). Nevertheless Table 6.2 shows the net growth of internal migration to rural units from 1975 and net migration from urban units (2,000+ inhabitants).

As already noted, these tendencies are minimised by the new INSEE geographical delimitation. Nevertheless, Table 6.3 reveals a quite negative correlation between the degree to which urban poles are urbanised and their population changes. Since 1975, urban emigration was high for all urban poles, especially those of over 20,000 in employment. Overall, net in-migration was

Table 6.2   Annual percentage change of rural population, 1954–90

		Natural change	Net migration	Total
Rural communes	1954–62	+0.51	-0.96	-0.45
	1962–68	+0.31	-0.75	-0.44
	1968–75	+0.02	-0.12	-0.12
	1975–82	-0.12	+0.99	+0.87
	1982–90	+0.02	+0.82	+0.84
Urban units	1982–90	+0.55	-0.16	+0.39

Source: SEGESA and INSEE 1992.

higher for INSEE 'dominant rural areas' than for 'dominant urban areas', although the latter incorporates growing peri-urban communes. Many rural areas distant from the commuting influence of the urban poles gained population, with remote rural areas almost attaining net in-migration by 1990. Though at a lower degree, these tendencies have been largely maintained in the last ten years and are confirmed by the first results of the 1999 population census, shown in Table 6.4. Even remote rural areas are now showing net in-migration, in spite of a strong natural decline due to their elderly age structures (see below).

    Although the reversal of long-term demographic rural trends is the consequence of positive net migration, little quantitative data is available on who these in-migrants are, in terms of their socio-economic and cultural background. Fortunately, evidence from regional and local case studies can help us generalise in time and space.[8] During the 1960s in-migrants were mostly retired people, either coming back to their birthplace at the end of their working life or those from an urban origin choosing rural communes and village houses because of lower costs and/or the natural and more human environment. Yet soon, in-migrants to rural areas were also workers and young people. They were attracted by rural firms and employment in the tertiary sector (old people's homes, centres for disabled people, services) but mostly by the desire and aspiration to live in and own a family house, which was only possible in the rural periphery of big towns (Berger et al., 1980). The choice of 'life location' (*lieu de vie*) became more important than that of 'working place' (*lieu de travail*). The extent that peri-urban rural communes adopted these 'peculiar' characteristics – young couples with children dominant, high numbers of white collar workers or even the working class – depended on local land values, the physical environment, the type and quantity

Table 6.3   France's demographic evolution, 1975–90

	Size of (influential) urban pole (no. in employment)	Annual change (%)		Due to net migration (%)		
		1975–82	1982–90	1968–75	1975–82	1982–90
Urban poles	5,000–19,999	+0.35	+0.24	+0.43	-0.20	-0.20
	20,000–99,999	+0.14	+0.18	+0.34	-0.57	-0.44
	100,000+	+0.20	+0.51	+0.26	-0. 44	-0.17
Peri-urban	5,000–19,999	+1.76	+1.20	+0.37	+1.69	+0.98
communes	20,000–99,999	+2.30	+1.51	+0.99	+2.04	+1.13
	100,000+	+2.84	+1.75	+2.08	+2.52	+1.83
Dominant urban area		*+0.59*	*+0.64*	*+0.46*	*+0.02*	*+0.07*
Rural under weak	5,000–19,999	+0.35	+0.39	-0.47	+0.52	+0.46
urban influence	20,000–99,999	+0.43	+0.54	-0.32	+0.55	+0.56
	100,000+	+0.69	+0.90	-0.15	+0.86	+0.94
Rural poles		+0.17	-0.11	+0.25	-0.17	-0.31
Periphery of rural poles		+0.48	+0.56	-0.55	+0.66	+0.58
Remote rural		-0.34	-0.31	-0.65	-0.03	-0.01
Dominant rural area		*+0.09*	*+0.11*	*-0.40*	*+0.23*	*+0.22*
Total France		**+0.47**	**+0.52**	**+0.24**	**+0.07**	**+0.10**

Source: INRA and INSEE, 1998, p. 33.

Table 6.4   France's demographic evolution, 1990–99

	Area (%)	Population 1999 (millions)	(%)	Internal change 1990–99 (%)		
				Annual change	Natural change	Net migration
Urban poles	7.4	35,155	60.1	+0.26	+0.55	-0.29
Peri-urban communes	22.1	9,664	16.6	+0.98	+0.38	+0.60
Dominant urban area	*29.5*	*44,819*	*76.7*	*+0.41*	*+0.51*	*-0.10*
Rural under weak urban influence	23.8	5,304	9.1	+0.52	-0.03	+0.55
Rural poles	2.5	2,110	3.6	+0.01	+0.02	-0.01
Periphery of rural poles	7.1	1,169	2.0	+0.24	+0.01	+0.24
Remote rural	37.2	5,036	8.6	-0.06	-0.34	+0.28
Dominant rural area	*70.6*	*13,619*	*23.3*	*+0.20*	*-0.13*	*+0.34*
Total France	**100**	**58,439**	**100**	**+0.36**	**+0.36**	**–**

Source: Piguet and Schmitt, 1999.

of housing, local power relations, the social composition of municipalities and, above all, accessibility (Berger, 1990; Berger and Saint-Gérand, 1999). Recently, an increasing in-migration of 'poor' people (redundant, unemployed, evicted, homeless) has even been noted (De Lafont, 1997). Added to the fact that local out-migration to urban areas continues, especially for higher education and by the highly educated, the rural social structure is becoming characterised by lower incomes, rather modest socioeconomic status (petite bourgeoisie) and even proletarianisation.

A Double Cause

Beyond this descriptive analysis of who in-migrants to rural areas are, we need a theoretical perspective to explain the end of the net rural exodus in France. Two kinds of economic and social factors seem crucial. First is the revalorization of countryside and agrarian and natural landscapes – a 'neo-ruralism' – in French post-industrial society. This began with a first wave of the anti-materialist *néo-ruraux* after 1968 (Léger and Hervieu, 1979), a very important movement for explaining the mental reverse concerning rural values. Countryside and peasantry, previously associated with archaism and despised, became suddenly attractive and synonymous with valuable areas for recreation and natural, modest ways of living. This signalled the start of an anti-urban social movement, the development of rural second homes and diverse people 'returning' to villages and rustic life. Of course, at the same time, agriculture and the countryside were involved in major modernisation. The consequences of the latter were in contradiction with what was called in French social research the 'invented countryside' and 'dreamed countryside' (for example, by Jean Viard) or 'rustic utopia' (for example, by Henri Mendras), instead of a 'rural idyll' as in Britain. In spite of the distance between 'degrading' rural reality and an ideal *ruralité*, moving to rural France persisted in the 1980s. Later, in the 1990s, as unemployment and the precariousness of economic activity increased, a new wave of migrants moved to rural areas – people made redundant, the unemployed, inactive young people, RMIstes[9] – all hoping to find better conditions of living, even if very modest (Balley, Lenormand and Mathieu, 1992; Mathieu 1995b, 1997). Recently some associations and municipalities have even mobilised to organise services to welcome (lodging) and integrate (cultural activities) this population of 'town refugees' (Epagneul and Mathieu, 2000).

The second key factor in rural 'recomposition' is located in the changing rural employment structure. The decline of employment in primary production was independent from the rising non-agricultural employment trend (Bontron and Mathieu, 1980; Mathieu, 1985). Whilst overall manufacturing employment had declined severely since 1974, heralding the general deindustrialisation of the French economy, small industrial enterprises were being created in rural areas and existing rural industries showed more resistance to decline. This was because of lower costs and salaries as well as labour qualities and rural amenities. Such a ruralisation of industry has been documented from the 1970s (Bontron and Mathieu, 1977; Mathieu and Vélard, 1981) to the 1990s (Mathieu, 1995b; Allaire, Hubert and Langlet, 1996). At the same time, construction and service employment was augmented in response to the arrival of new rural inhabitants and the restoration of farms and rural buildings. Table 6.5 illustrates the growth of non-agricultural employment in two types of rural space (outside ZPIU and inside ZPIU), especially in the latter. This percentage augmentation is higher than the percentage of total employment growth of urban units.

Table 6.5   Employment profiles based on 1982 area delimitations (thousands)

	Total 1982	1982–90	% change
Agricultural and related employment	952	-262	-27.5
Non-agricultural employment	732.1	+39	+5.3
Rural outside ZPIU	*1,684.1*	*-223*	*-13.1*
Agricultural and related employment	626	-141	-22.6
Non-agricultural employment	1,677	+204	+12.1
Rural in ZPIU	*2,303*	*+63*	*+2.7*
Total rural communes	*3,987*	*-160*	*-4.0*
Urban units<20,000 inhabitants	3,436	+120	+3.5
20,000-200,000 inhabitants	5,082	+184	+3.6
Total France	**21,357**	**+702**	**+3.2**

Source: General Census of Population 1982 and 1990 (cf. Khaldun, Mathieu and Rémy, 1999).

To this quantitative perspective, more qualitative insights must be added. Many studies indicate the role of private and familial strategies (Mathieu, 1995b). Facing increasing unemployment, even in urban areas, strategies developed new employment by mobilising local resources and the advantages of a better 'quality of life' in rural areas. Part-time employment was no longer despised but accepted as a means of creating new jobs and enterprises and of saving the reproduction of the family. This tendency reveals a social reflexivity concerning work in post-industrial society and explains the development of initiatives and innovation at a micro-level. Even on farms, new ideologies of local resources and sustainable agriculture and new representations of work could lead to the maintenance and even the creation of employment (for example, pluri-activity, diversification, direct sales, tourism).

A Selective Geography

There is, of course, a spatial differentiation of this 'rural renewal' or 'counter-urbanisation'. The net migration dimension and different waves of neo-ruralism were the most powerful and effective in the south of France. For example, the post-1968 *néo-ruraux* were associated with places such as Cévennes, Ardèche and Limousin. Indeed, neo-ruralism has also been labelled a 'sun migration' movement (Rouzier, 1990). Some inquiries even mentioned industrial workers' families moving from regions in crisis (for example, the coal, textile and steel basins of northern France such as Lorraine, Nord-Pas-de-Calais) to rural Mediterranean *arrière pays* villages or little towns, even if they did not have employment. This represents a collective social representation emerging strongly through the 1990s based on the idea that a modest way of life or even poverty is more supportable in the countryside than in the big cities. Such trends worsened the differentiation between rural and urban areas' incomes.

The early counterurbanisation tendency continues to influence peri-urban areas (see Table 6.3) and especially Paris, Lyon and the regional metropolises. The taste for individual houses and a village environment continued amongst the upper class and extended to the petite bourgeoisie, with a multiplication of rural housing estates and a generalisation of this type of geographical mobility. As suggested in Tables 6.3 and 6.4, rural areas distant from metropolitan areas have been touched by this taste for 'nature' and 'natural' spaces and landscapes. Even the remote rural area of northern France (except the crisis areas) registered in-migration, though at lower levels, based on this rustic ideology. Rural areas in decline are now restricted not to those where agricultural activity is still important but just to where industries are in such a crisis that every administrative regional unit, urban and rural, is declining through emigration and employment loss.

The Administrative Structure of Rural France: Between Fragmentation and Centralisation

France is well known in Europe as having the most dispersed administrative structures governing the management of rural areas, whilst at the same time having the most centralised state, especially with regard to agricultural activity. There is doubt as to whether the decentralisation law initiated by President Mitterand in 1982, which aimed to build up regional power at the expense of the local level, completely fulfilled its targets.

Rural France: A Multitude of Small Municipalities

Rural France is administrated by a spatial structure inherited from the Revolution, guided by a principle of spatial equity and accessibility. Deeply anchored in French political culture (Pinchemel, 1981), this principle explains the hierarchical division of territory into communes, cantons and departments, with their different levels of county town (*chef-lieu*), and the persistence of small basic institutional units such as rural communes. In 1990 there were 31,251 rural municipalities (0–2,000 inhabitants), containing 14,717,396 inhabitants, and just 5,300 urban municipalities (2,000+ inhabitants), holding the remaining 74 per cent of French population. In the first category, Table 6.6 gives an idea how small (in population) rural communes are, yet all have elected mayors and municipal councillors (their number depending on their population over a 500-inhabitant threshold).

Table 6.6   Population of rural and urban communes, 1990

Size	Communes		Population		Area	
	No.	%	No.	%	Km^2	%
<50 inhabitants	1,111	3.0	38,115	0.1	7,815	1.4
50–99	3,016	8.3	226,159	0.4	25,450	4.7
100–199	6,728	18.4	991,459	1.8	70,991	13.1
200–499	10,620	29.1	3,413,187	6.0	146,023	26.8
500–999	6,036	16.5	4,169,280	7.4	108,005	19.9
1,000–1,999	3,064	8.4	4,177,993	7.4	72,047	13.2
2,000+	676	1.8	1,701,203	3.0	23,985	4.4
Rural total	31,251	85.5	14,717,396	26.0	454,316	83.5
Urban total	5,300	14.5	41,897,759	74.0	89,649	16.5
Total France	36,551	100	56,615,155	100	543,965	10

Source: INSEE, 1990.

Rural mayors have some local responsibilities, notably dealing with public order and local roads and paths. They have more of a role in local development through their authority in granting building licenses (*permis de construire*) and in their decision-making powers over managing services and equipment (primary schools, water equipment, local management and planning), even if these tasks are also linked with cantons (elected *conseiller général*) and departmental authorities (elected *conseil général* assembly, the préfecture representing the central state). This situation gives a chance for development to very small units if the mayor is a political or economic leader. This has been noted, for example, in Corrèze (Mathieu, 1986a), where young Jacques Chirac began his (brilliant) career by providing health services in two conflicting communes of the same canton.

The limited local power of very small communes can, however, be considered a handicap, linked to weak financial resources that are insufficient to provide territorial management. From the central government point of view, the multitude of local structures weakens comprehensive rural development, as well as compromising France's position in the EU. From the 1970s to the 1990s, many laws tried to reduce the number of communes, inviting them first to merge, then to group into syndicates and then to collaborate financially in projects of *intercommunalité*. Such policies had very little success (the number of communes today remains nearly the same as after the Revolution). Communes are still considered as the first level of democracy and political expression. There has been some grouping of communes but principally these are big towns (*communautés urbaines*) grouping for financial and equipment management. Such groupings are also present in rural areas but they are also limited to economic management (refuse collection, tourism projects, services and equipment). A new level of administration is emerging but, though gathering communal representatives, their political responsibility is not recognised (or understood) by local people.

Agricultural Activity: A Strong Centralised Territorial Administration

Alternatively, the administration of rural areas can be seen as managed by central government, and by the Ministry of Agriculture and Forestry in particular. There is a very strong tradition within the French government to consider food production and the peasantry as quasi-state 'properties'. Consequently, farmers' disputes and claims are always directed against the state (Plet, 1993). From the 1950s until the first CAP reforms, it was accepted that the central state and the 'official' agricultural unions (FNSEA and CNJA[10]) jointly managed farmers, agriculture and agri-industries in a corporatist direct relation called 'cogestion' (joint management). Even when the necessity of regional policies and orientation were recognised, resultant policies (for mountains and less favoured areas) could only be elaborated by the Agriculture Ministry and specific institutions created for their application. Contradicting the strong tendency for a diversification of local development, this centralised policy structure promoted the homogenisation of rural areas, as in the Bretagne region with its agricultural modernisation. Even the decentralisation process initiated in 1982 by President Mitterand, consolidating 21 metropolitan regions by giving them economic, educational and environmental responsibilities, did not break this quasi-monopoly on the primary sector and the prospective development of rural areas. Only recently, the 1999 Law of Agricultural Adjustment (Loi d'Orientation Agricole) displayed continued evidence of the direct link between central government and individual farmers. For farm contracts, each farmer must negotiate with the state authority itself, through the relevant departmental commissions.

Defining and Controlling Rural Development Policies: A Bottom-up/ Top-down Conjunction

'Local Development' or the 'Country Movement'

Probably because of the contradictions between local decentralised power and centralised public policy, initial experiments in local development came out of rural areas as *développement de pays* (country development). During the 1960s, this dynamic began around regionalist leaders, such as Paul Houée in Le Ménée in Brittany (Houée, 1972, 1987, 1989, 1992), and peasant leaders, such as De Cafarelli in Thiérache in the north (Bonerandi, 1999). This dynamic was based on a common ideology that modernisation and social progress could not be achieved if cultural and territorial factors were not taken in account. Local development was defined in terms of a mobilisation of local social forces (forces vives, living forces) in a collective way, uniting elected leaders and voluntary actors in a belief in the capacity of a territory to reverse the general tendency of rural and agricultural decline.

In their extension, these bottom-up local or 'country' movements reflected the parallel territorial and rural policies which were being built into different public spheres. These aimed for a top-down local restructuring of rural areas in crisis. Born during the last war, a counter-centralisation ideology pointed out the negative economic and social consequences of Paris and its region's continual growth. This ideology first penetrated the Commissariat Général du Plan, then the Ministry of Agriculture in the 1960s, and finally in the 1970s DATAR (Délégation à l'Aménagement du Territoire et à l'Action Régionale), the inter-ministerial committee supposed to manage France's territorial equilibrium and development. A local/global conception of regional planning became the rule. Different instruments were applied, such as Rural Management Plans (Plan d'Aménagement Rural), national and regional Countryside Contracts (Contrats de Pays) and even Regional Natural Parks (Parcs Naturels Regionaux). All reflected the same ideology: a resistance to crisis and regressive demographic and economic trends through mobilising local forces and powers in territorially-anchored development (Bontron et al., 1983).

Figure 6.1 shows the extent of local development planning just before European programmes for rural development were promoted. This French rural policy mixed innovation coming from local populations with that coming from central, regional and local leaders, both elected (mayors, conseillers généraux, deputies) and nominated (préfet and sous-préfet, departmental administration, Regional Park directors). In this process, one must note the role of professional local development activity leaders, the 'agent de développement' and the 'animateur de pays', paid through local associations subsidised by national, regional and local administrations. Consequently, if gradually, some rural 'pays' became archetypal examples of local development, since they used in succession every procedure (from Plan d'Aménagement Rural to inter-communal syndicates) in order to consolidate social and economic growth (for example, Pays de Die, Larzac).

There is an important literature analysing the country movement in France and it remains a significant field of investigation. Publications include those of authors trying to discuss if local development was a fashion or a social movement (Jollivet, 1985; Pecqueur, 1989), papers analysing the respective roles of state authorities and local elites (Mathieu and Mengin, 1988), authors trying to evaluate the territorial effect of local development on local society (Barthe, 1992, 1998; Mathieu, 1989, 1995c), and studies seeking to compare the dimensions of and actors involved in local development in urban and rural contexts (Mathieu, 1986b; Mathieu and Robert, 1998).

European Programmes

Objective 5b of the European Structural funds and the LEADER programmes did not introduce a discontinuity within French rural development policy seen as a whole. The idea of funding 'disadvantaged rural areas' had been at the core of management policy at a national level since the 1960s, and European

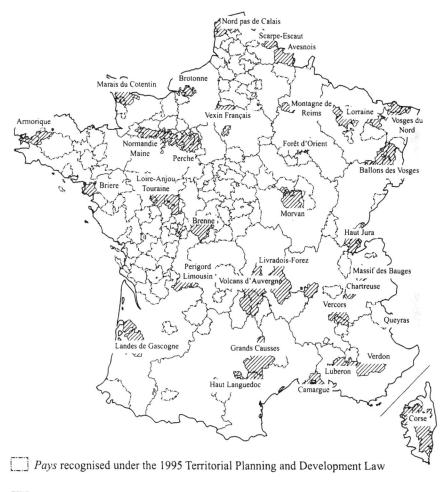

[ ] *Pays* recognised under the 1995 Territorial Planning and Development Law

[//] Regional Natural Parks

Figure 6.1   Designated countryside and Regional Natural Parks, 1999

Source: ETD, 2000.

policy was able to use the delimitation of 'cantons fragiles' (fragile cantons) established by SEGESA/DATAR at the end of the 1980s (Bontron and Aitchison, 1984; SEGESA, 1992). Objective 5b funds have only been important in introducing more strongly rural development as a priority at the regional level, giving decentralisation an added significance. Of course, where rural policies and local initiatives existed, EU funds reinforced their orientations, and, where they did not exist or looked weak, they opened up

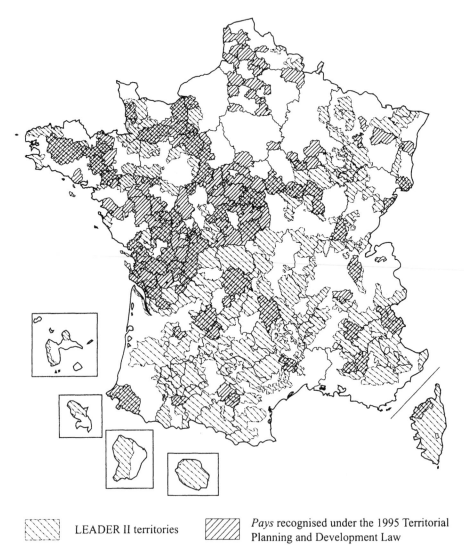

| LEADER II territories | | *Pays* recognised under the 1995 Territorial Planning and Development Law |

Figure 6.2   Designated countryside and LEADER territories, 1999

Source: DATAR, CNASEA/SEGESA 1999.

the process of defining a rural area, its problems and possible solutions to some new localities or regional authorities. A serious evaluation of the European Structural fund policy is required before we can draw conclusions about their impact on disadvantaged rural areas untouched by earlier French rural development planning.

The same hypothesis can be advanced for the potential impact of LEADER programmes. These programmes, answering to the new EU target of fitting policies to territories instead fitting territories to procedures, opened new opportunities and margins for local actors to consolidate their developing abilities (Barthe, 1998). An assimilation between 'pays development' and the 'territorial project of development' has been confirmed. However, it is very difficult to evaluate the reality and the type of development associated with those programmes, as well as their relations with two kinds of local development definition: the administrative, aiming to simplify the 36,000 communes, and the developmental, aiming to solve the crisis tendencies (rural unemployment, lack of services and equipment, environmental problems). When looking at Figure 6.2, showing LEADER II territories and *pays* recognised under the 1995 Territorial Planning and Development Law (Loi d'Orientation pour l'Aménagement et le Développement du Territoire), and when comparing it to Figure 6.1, several observations can be made. First we notice differences in the size and coverage of the territories and their increase over time. It suggests an appropriation of European programmes to build new areas of management fitting with the European geographical conception of regional and local policies. Secondly, except in western France, LEADER programmes scarcely coincided with nationally recognised *pays*. The designations look complementary, showing most congruence with older local rural development areas, such as Regional Natural Parks. Though we need more studies to validate this hypothesis, it seems that LEADER and the European rural development programme have probably modified leaders' and social groups' landscape without strongly changing the rural geography of power and empowerment. They also probably have not reduced the French gap between urban and rural development policies.

Discourses of Urbanity and Rurality and the Territorial Development Process in France

As noted at the start of this chapter, 'the rural' remains a concept intensely mobilised in every sphere of French society: by ordinary people; social movements and associations; local, regional and central policies; and of course, by the social sciences. For ordinary people, in almost all social classes, *la campagne* (countryside) appeals as a positive representation (Hervieu and Viard 1996), linked with physical properties (pure air, natural environment, smells, landscapes) but also with moral qualities (liberty, conviviality, a specific 'village' sociability, local management, democracy). Social movements, syndicates and associations are also mobilised by the 'rural cause'. This is most obvious in farmers' movements. Conservatives, such as the FNSEA or the CNJA (see note 10), fight against 'desertification' and have recently claimed an interest in protecting the environment and natural resources (cf. the Farre network for a 'rational' agriculture). Radicals, such as the Confédération Paysanne (Peasant Confederation), struggle against globalisation's consequences for food and rural employment and for the challenge of 'sustainable' or 'peasant' agriculture (Confédération Paysanne, 1994). Rural or 'local' identity is more and more important inside the developing associations movement. This phenomenon has been recognised among both established groups, such as the MRJC (Rural Movement of Christian Youth) (Epagneul and Mathieu, 2000), and newer groups, such as local associations concerned with re-integrating the unemployed and socially excluded. It is also apparent in big networks of rural associations, such as UNADEL (Union of National and Local Development Associations) or CELAVAL (Centre for the Liaison of Associations Valorising the Rural Milieu). Though many ministries (the Treasury, for example) show hostility to specific rural and local policies, public policies are still influenced by the issues of rural development, landscape protection, natural and environmental problems, local and 'country' development and, as said before, agriculture. Discussions in parliament and resulting legislation are there to prove it. Rural discourses and representations shape, at many levels, French social and spatial change.

Beyond this general overview, it is possible to pick up some features more specifically French.[11] In France, 'the rural' has always been understood in conjunction with 'the urban' and their inter-relationship causes persistent controversy. Is there a 'wasteland France' (Fottorino, 1989), a 'France for departures' (Alphandéry, Bitoun and Dupont, 1989) and a 'rural crisis' (Béteille, 1994), or a 'rural renaissance' and a 'chosen countryside' (Kayser, 1990, 1996)? Is there an urban crisis linked with a rejected style of life (Mathieu, 1996), or is this crisis more of a consequent negative effect which comes from paying too much attention to rural problems and rural areas (Lévy, 1994)? Is it the end of the rural/urban opposition or the emergence of new relationships based on differentiation and complementarity?

Rural and urban social representations, as well as their inter-relation, have moved in time and space (Mathieu, 1990, 1998; Mathieu and Robert, 1998). Analysing these variations from the 1950s to the 1990s, periods can be distinguished based on the dominant ideology of each period. There are very subtle links between both main ideologies: rural disappearance ('the rural death') and urban triumph ('general urbanisation'). These links depend on the periodical social construction of the definition and contents of what is rural and urban, and their comparative values. In the 1960s, urbanity and an urban style of life were considered modern and progressive in every sphere of French society except among agricultural and rural movements. The latter's fight against this dominant representation involved providing a new definition of ruralité as a type of space characterised by non-agricultural activities, diversification and rural industries. In the 1990s, 'rurality' became synonymous with 'nature', sustainable activity, landscape and well-being. A post-industrial (postmodern?) conception dominated, and urbanity became a very controversial concept. Valorised by adherents of an equivalence between urbanity and citizenship, it became for others synonymous with unhealthy, anti-natural and unsustainable lifestyles. A reactivation of the rural/urban antagonism is emerging, based on differences in environmental qualities of life.

A rural/urban relationship observatory (Mathieu and Robert, 1998) has been created to clarify the complex relations between each ideological stream and social class. Its main aim is to explore the effectiveness of those social representations on decision-making and social practices. Recent and current case studies comparing urban and rural localities underline the important part played by ideologies and representations of rurality, urbanity, and the rural/urban relationship, on individual choices of 'home' location (often consolidating a nomadic style of life between countryside home and city home) and collective initiatives and territorial development policies. Coming from ruralité/urbanity discourses, the 'country' pattern of leadership and local power born in rural areas is consolidating itself. It penetrates the conception of towns' governance, stressing the necessity of rethinking urban areas (Blanc and Mathieu, 1996) through the dimension of their inhabitants and the way they perceive and act within different localities and 'milieu' ('life spaces' defined by interaction between material and social processes). There could be an emerging pattern of leadership and local power reconciling conceptions and theories, taking into account both the specificity of rural and urban milieu and their complementarity and systemic interaction, calling for solidarity in territorial policies and local governance.

Conclusion

Issues such as the influence of social structure, rural development policies and social representations on leadership and local power in rural France are clearly complex. Around the question of *ruralité*, there is an intricate interaction between the factual level (the 'real' of actions, decisions, practices) and the mental level (the 'ideal' of dreams, representations, discourses, desires). It is difficult to make out a hierarchy of causes, or to distinguish effects from causes. We have tried in this chapter to discuss key issues implicated in this question in France. Past and inherited structures lie heavy and must be taken into account to understand subsequent transformations in rural areas. Our thesis is that EU policies' efficacy often only confirms those inherited features and functioning mechanisms. Nonetheless, emerging attitudes and new 'ruralian'[12] ideologies are assuming greater significance, influencing choices and practices in many fields (residential, eating, working, moving, leisure and culture, socialising). We are in the vanguard of a rising social representation calling for a new right: to live a 'peasant' or 'rural' way of life even in urban areas, requesting a life-style choice characterised by a double identity, with both an urban and a rural belonging. If they care for their reputation, re-election and power, rural elites must take this new life-style into account.

Notes

1   The events surrounding the trial of the radical French farmer José Bové in Larzac on 30 June 2000 demonstrate the contemporaneity of this imagination. Two explanations can be advanced to explain this endurance. Firstly, at the state level, agriculture was an important preoccupation of both the royal and then the republican conceptions of the nation. Food autonomy and control of the primary sector were essential (cf. the establishment of the first national statistics at the beginning of the nineteenth century; the creation of a Ministry in the mid-nineteenth century and its prominent place in the ministries' hierarchy). Today, whilst liberalism and EU policy go against this representation, it is still traceable in agricultural policy (cf. 1999 legislation, the Loi d'Orientation Agricole, which both reiterates a successful 1960 law and gives a French angle to CAP policies by integrating environmental, landscape management and employment care within agricultural policy). Secondly, French society still valorises the peasantry and a peasant way of life, even though it is a reinvented rurality. This is discussed later in the chapter.

2   This term is still used by different social actors, such as the state, rural syndicates and associations, media, scientific literature and ordinary people. It is always used to designate a catastrophic demographic tendency that leads to rural areas becoming abandoned wastelands. The term 'desertification' suggests an unavoidable process. Analysis of this representation and the role it has played in the political sphere can be found in Mathieu and Duboscq (1985).

3   The Institut National de la Statistique et des Etudes Economiques is the French national statistical institute created after the Second World War. It plays a major survey role for the French government.

4   ZPIU was an INSEE statistical delimitation used until the 1990 census. It was created to distinguish rural communes with high levels of commuters and non-agricultural employment. For INSEE, both characteristics gave a commune and its periphery an urban status (ZPIU) even if small and devoted to 'rural' industries. Communes of less than 2,000 inhabitants (the official definition of a rural commune) could be in ZPIU. This definition was abandoned because in the 1990 census almost 80 per cent of France was in ZPIU and the 'hors ZPIU' (out of ZPIU) rural communes were ridiculously few in number and area.

5   Mathieu (1995a) has pointed out this relationship for the *rural profond* ('deep rural areas'), the term used before *rural isolé* ('remote rural areas').

6   For the same period of time (1982–90), INSEE's change of definition resulted in 14,717,396 inhabitants in 31,251 rural communes becoming 13,376,000 inhabitants in 21,766 rural communes plus 947 urban units comprising the new 'Dominant rural areas' (espace à dominante rurale).

7   SEGESA stands for Société d'Etudes Géographiques et Sociologiques Appliquées (created in 1967).

8   So many geographical rural studies at different levels have been undertaken in the universities that we cannot reference them explicitly. See the work of Jean Renard (University of Nantes), Bernard Kayser and Jean Pilleboue (University of Toulouse), Pierre Duboscq (University of Bordeaux), Marie-Claude Maurel and Marie Claire Bernard (University of Montpellier, where also work economists such as A. Berger and J. Rouzier) and, of course, the Parisian rural teams (GRS of Nanterre, SEGESA, STRATES of Paris 1).

9   The Revenu Minimum d'Insertion (RMI) is a financial public allocation to help economically marginalised people re-enter society.

10 The Fédération Nationale des Syndicats d'Exploitants Agricoles (National Federation of Farmers' Unions) and its Young Farmers'Association.

11 There is probably overlap here with other countries in Europe, such as Belgium (Mormont 1996, 1997).

12 We cannot say 'agrarian' because what is emerging is something different and not conservative.

References

Allaire, G., Hubert, B. and Langlet, A. (eds) (1996), *Nouvelles Fonctions de l'Agriculture et de l'Espace Rural, Enjeux et Défis Identifiés par la Recherche*, Institut National de la Recherche Agronomique, Toulouse.

Alphandéry, P., Bitoun, P. and Dupont, Y. (1989), *Les Champs du Départ, une France Rurale sans Paysans?*, La Découverte, Paris.

Balley, C., Lenormand, P. and Mathieu, N. (1992), 'Territoire Rural, RMI, Pauvreté', *Sociétés Contemporaines*, vol. 9, pp. 53–76.

Barthe, L. (1992), *Le Réquistanais: Émergence d'un Pays*, Université Paul Valery, Montpellier, Espaces et Sociétés en Transition, Etudes 1.

Barthe, L. (1998), *Processus de Différenciation des Espaces Ruraux et Politiques de Développement Local*, Doctoral Thesis, Université de Toulouse le Mirail.

Berger, M. (1990), 'A Propos des Choix Résidentiels des Espaces Périurbains: Peut-on Parler de Stratégies Territoriales?', *Strates*, vol. 5, pp. 75–84.

Berger, M., Fruit, J.P., Plet, F. and Robic, M.C. (1980), 'Rurbanisation et Analyse des Espaces Ruraux', *L'Espace Géographique*, vol. 9, pp. 303–13.

Berger, M. and Saint-Gérand, T. (1999), 'Entre Ville et Campagne, la Mobilité des Périurbains', *Observatoire des Rapports Entre Rural et Urbain*, vol. 2, pp. 1–4.

Béteille, R. (1994), *La Crise Rurale*, PUF, Paris.

Blanc, N. and Mathieu, N. (1996), 'Repenser l'Effacement de la Nature dans la Ville', *Le Courrier du Centre National de la Recherche Scientifique*, vol. 82, pp. 105-7.

Bonerandi, E. (1999), *Devenir des Espaces Ruraux et Élus Locaux: l'Exemple de la Thiérache*, doctoral thesis, Université de Paris I.

Bontron, J.C. (1983–84) 'Un Renversement de Tendances, la Campagne se Repeuple', *Aménagement et Nature*, vol. 72, pp. 21–2.

Bontron, J.C. (1989), 'De l'Exode Rural à l'Exode Urbain', in INRA and SEGESA [Institut National de la Recherche Agronomique and Société d'Etudes Géographiques et Sociologiques Appliquées] (eds), *Grand Atlas de la France Rural*, J-P. de Monza, Paris, pp. 44–5.

Bontron, J.C. and Aitchison, J. (1984), 'Les Zones Rurales Fragiles en France. Une Approche Méthodologique', *Bulletin de la Société Neuchâteloise de Géographie*, vol. 28, pp. 23–53.

Bontron, J.C., Gilette, Ch., Mathieu, N., Peyon, J.P., Plet, F. and Robic, M.C. (1983), 'Eléments de Réflexion sur l'Aménagement et l'Espace Rural en France', *Geographia Polonica*, vol. 46, pp. 193–216.

Bontron, J.C. and Mathieu, N. (1968), 'Repenser l'Espace Rural', *Paysans*, vol. 70, pp. 99–107.

Bontron, J.C. and Mathieu, N. (1973), 'Les Transformations de l'Espace Rural. Problèmes de Méthode', *Etudes Rurales*, vol. 58–59, pp. 137–59.

Bontron, J.C. and Mathieu, N. (1977), *Industrie et Espace Rural: les Apports d'une Analyse Bibliographique*, SEGESA/Acear, Paris.

Bontron, J.C. and Mathieu, N. (1980), 'Transformations Agricoles et Transformations Rurales en France Depuis 1950', *Economie Rurale*, vol. 137, pp. 3–10.

Confédération Paysanne (1994), *L'Agriculture Paysanne: des Pratiques aux Enjeux de Société*, FPH/Confédération Paysanne, Paris.

De Lafont, V. (ed.) (1997), *L'Exclusion en Milieu Rural. Synthèse des Travaux*, Commissariat Général du Plan, Paris.

Epagneul, M-F. and Mathieu N. (2000), 'Explorer le Rôle des Associations dans l'Insertion et la Création d'Emploi en Milieu rural', *Economie Rurale*, vol. 258.

Fottorino, E. (1989), *La France en Friche*, Lieu Commun, Paris.

Hervieu, B. and Viard, J. (1996), *Au Bonheur des Campagnes*, Editions de L'Aube, Paris.

Houée, P. (1972) *Les Étapes du Développement Rural*, Editions Ouvrières, Paris.

Houée, P. (1987), 'Démarches de Développement Local en Milieu Rural: l'Expérience du Méné (1965–1986)', *Les Cahiers de la Recherche-Développement*, vol. 13, pp. 5–11.

Houée, P. (1989), *Les Politiques de Développement Rural: des Années de Croissance aux Temps d'Incertitude*, Institut National de la Recherche Agronomique – Economica, Paris.

Houée, P. (1992), *La Décentralisation: Territoires Ruraux et Développement*, Syros, Paris.

INRA and INSEE [Institut National de la Recherche Agronomique and Institut National de la Statistique et des Etudes Economiques] (1998), *Les Campagnes et Leurs Villes. Contours et Caractères*, INSEE, Paris.

INSEE (1990), *Recensement de la Population de 1990*, INSEE, Paris.

Jollivet, M. (1985), 'Le Développement Local, Mode ou Mouvement Social', *Economie Rurale*, vol. 166, pp. 10–6.

Jollivet, M. (ed.) (1997), *Vers un Rural Postindustriel. Rural et Environnement dans Huit Pays Européens*, l'Harmattan, Paris.

Kayser, B. (1990), *La Renaissance Rurale, Sociologie des Campagnes du Monde Occidental*, Armand Colin, Paris.

Kayser, B. (1996), *Ils ont Choisi la Campagne*, Editions de l'Aube, Paris.

Khaldun, I., Mathieu, N. and Rémy, J. (1999), 'Travaux Ruraux: Vers un Nouveau Contrat Social?', *Témoin, Défense et Illustration du Monde Rural*, vol. 15, pp. 105–15.

Léger, D. and Hervieu, B. (1979), *Le Retour à la Nature. Au Fond de la Forêt, l'Etat*, Le Seuil, Paris.

Lévy, J. (1994), 'Oser le Désert? Des Pays sans Paysans', *Sciences Humaines*, vol. 4, pp. 7–8.

Mathieu, N. (1974), 'Propos Critiques sur l'Urbanisation des Campagnes', *Espaces et Sociétés*, vol. 12, pp. 71–89.

Mathieu, N. (1985), 'Un Nouveau Modèle d'Analyse des Transformations Encours: la Diversification-Spécialisation de l'Espace Rural Français', *Economie Rurale*, vol. 166, pp. 38–44.

Mathieu, N. (1986a), 'Le Limousin', in Y. Lacoste (ed.), *Géopolitiques des Régions Françaises. II. La Façade Occidentale*, Fayard, Paris, pp. 847–941.

Mathieu, N. (1986b), 'Les Dimensions et les Acteurs du Développement Local: Éléments pour une Analyse Comparée Urbain/Rural', *Les Cahiers de Fontenay*, vol. 41/42/43, pp. 81–92.

Mathieu, N. (1989), 'Solidarité, Identité, Innovation, les Tensions Fondatrices de la Société Méjanaise', *Annales du Parc National des Cévennes*, vol. 4, pp. 229–61.

Mathieu, N. (1990), 'La Notion de Rural et les Rapports Ville/Campagne en France, des Années Cinquante aux Années Quatre Vingt', *Economie Rurale*, vol. 197, pp. 35–41.

Mathieu, N. (1995a), 'La Notion de Rural Profond, à la Recherche d'un Sens', in R. Béteille and S. Montagné-Villette (eds), *Le 'Rural Profond' Français*, SEDES, Paris, pp. 115–22.

Mathieu, N. (ed) (1995b), *L'Emploi Rural, une Vitalité Cachée*, l'Harmattan, Paris.

Mathieu, N. (1995c), 'Les Nouveaux Enjeux d'Appropriation et d'Usages des Causses: que Dire Depuis le Causse Méjan?', in J.L. Bonniol and A. Saussol (eds), *Grands Causses, Nouveaux Enjeux, Nouveaux Regards*, Presses de Causses Cévennes, Millau, pp. 357–67.

Mathieu, N. (1996), 'Rural et Urbain: Unité et Diversité dans les Évolutions des Modes d'Habiter', in M. Jollivet and N. Eizner (eds), *L'Europe et ses Campagnes*, Presses de Sciences Po, Paris, pp. 187–216.

Mathieu, N. (1997), 'Les Enjeux de l'Approche Géographique de l'Exclusion Sociale', *Economie Rurale*, vol. 224, pp. 21–8.

Mathieu, N. (1998), 'La Notion de Rural et les Rapports Ville/Campagne en France. Les Années Quatre-Vingt-dix', *Economie Rurale*, vol. 247, pp. 31–8.

Mathieu, N. and Duboscq, P. (1985), *Voyage en France par les Pays de Faible Densité*, Editions du CNRS, Toulouse.

Mathieu, N. and Mengin, J. (1988), 'Les Politiques de Développement Rural: Unité ou Diversité', in M. Jollivet (ed.), *Pour une Agriculture Diversifiée*, l'Harmattan, Paris, pp. 268–82.

Mathieu, N. and Robert, M. (1998), 'Pourquoi un Observatoire des Rapports Urbain/Rural?', *Observatoire des Rapports Entre Rural et Urbain*, vol. 1, pp. 1–4.

Mathieu, N. and Vélard, L. (1981), *L'Emploi et les Activités en Milieu Rural. Spécificités et Tendances d'Evolution. Une Approche Sectorielle et Régionale*, SEGESA, Paris.

Mormont, M. (1996), 'Le Rural Comme Catégorie de Lecture du Social', in M. Jollivet and N. Eizner (eds), *L'Europe et Ses Campagnes*, Presses de Sciences Po, Paris, pp. 161–76.

Mormont, M. (1997), 'A la Recherche des Spécificités Rurales', in M. Jollivet (ed.), *Vers un Rural Postindustriel. Rural et Environnement dans Huit Pays Européens*, l'Harmattan, Paris, pp. 17–44.

Pecqueur, B. (1989), *Le Développement Local, Mode ou Modèle?*, Syros, Paris.

Piguet, V. and Schmitt, B. (1999), *Premiers Résultats du RP 1999 Selon le Zonage en Aires Urbaines et son Complément Rural*, INRA-ENESAD, Dijon.

Pinchemel, P. (ed.) (1981), *La France, Activités, Milieu Ruraux et Urbains. Tome 2*, Armand Colin, Paris.

Plet, F. (1993), 'Colères Paysannes Face à l'Etat', *Hérodote*, vol. 69–70, pp. 170–84.

Rouzier, J. (1990), 'La Mutation de l'Arrière pays Méditerranéen ou un Modèle pour la Revitalisation des Communes Rurales', *Revue d'Economie Régionale et Urbaine*, vol. 5, pp. 695–713.

SEGESA (Société d'Etudes Géographiques et Sociologiques Appliquées) (1990), *Les Dynamiques Récentes de Création et de Localisation des Activités en Zone Rurale*, DATAR, Paris.

SEGESA (1992), *La Recomposition du Territoire. Essai de Typologie Économique des Cantons Français*, DATAR, Paris.

Chapter Seven

Recent Rural Restructuring in East and West Germany: Experiences and Backgrounds

Lutz Laschewski, Parto Teherani-Krönner and Titus Bahner

Introduction

Germany today is marked by the huge impacts of the post-war era, in which the country was divided into two states fronting opposing political blocs. Until 1989, the concept of rurality had a different meaning in each part of Germany and was linked to contrasting political objectives and sets of policy measures. As a result, we find huge structural differences between former East and West Germany, which challenge the unification project. Ten years after this unification in October 1990, East–West differences in every indicator overlay all other differences. Furthermore, there are also other historical boundaries of significance for rural development, in particular because they have affected and still influence land ownership and farm structures. In both parts of Germany we find also a north–south and, less significantly, a west–east divide (Struff, 1997). The most important cultural division is between Protestants and Catholics, the predominantly north–south distinction manifested in the Westphalia Peace of 1648. A second dividing line is the river Elbe, where regions to the east were Germanised during medieval colonisation. Large, often feudal, farm estates dominated the farm structure of these regions. Today, prosperous regions are more likely to be found in the southwest, while regions with structural problems are mostly found in the northeast.

Unification meant the transfer of institutions from West to East which had evolved under the particular conditions of post-war West Germany. This transfer allows us, for the purpose of this chapter, to describe only one institutional framework. However, we must note that these institutions do not necessarily work similarly in the East as in the West. Occasionally, therefore, we address issues of 'institutional fit' in the former East Germany. However, even if not explicitly mentioned, East–West differences affect all issues discussed in this chapter. Furthermore, this political framework of regional and agricultural policies has come under pressure after unification, not only because of a growing heterogeneity between regions but also because of the increase in federal states, each participating in decision-making processes, and the changed geopolitical context.

Socio-demographic Changes and Economic Restructuring in Rural Germany

Population in Rural Areas

Following the OECD definition, only a fifth of German territory is rural, and only eight per cent of the population lives in rural regions. Population density in German rural areas is higher than in most other countries, and differences between urban and rural regions are lower (Schrader, 1999). Nonetheless, in contrast to the OECD, the official German definition of rural areas,[1] which is used in subsequent discussion, covers a much wider area. The share of the total population living in rural areas is much higher in East than in West Germany, at 30 per cent and 18 per cent, respectively. This is shown in more detail in Table 7.1.

Table 7.1   The significance of rural areas in Germany

| | | | Type of rural region | | |
			Type I (agglomerations)	Type II (urbanised)	Type III (rural)
Population share (%)	West	100	2.9	7.1	7.7
	East	100	8.5	10.0	11.5
Share of land (%)	West	100	6.5	16.3	20.7
	East	100	13.0	18.1	35.2
Population density	West	254	115	111	95
(persons/ km^2)	East	164	108	91	54

Source: BflR, 1995; data for rural Kreise as defined in footnote 1.

The comparatively high share of rural Kreise (counties) in East Germany which are close to agglomerations is also a result of a high concentration of the population in the larger cities (Kroner, 1993). During the 1980s, West

Germany and East Germany showed different migration trends. While East Germany was characterised by a continuing and strong urbanisation, the development of West Germany was more of a retarded urbanisation (Kontuly and Dearden, 1998). The reasons for the latter were similar to those used to explain counterurbanisation processes in countries such as the United Kingdom, France and Italy: decentralisation of the labour force, house price differentials, changing demographic structures and policy factors. It might only be due to the comparatively homogenous settlement structure that true counterurbanisation did not take place in West Germany. Urbanisation in East Germany was the result of housing policy, with mobility taking place because administrative restrictions were very limited.

Unification had a significant impact on mobility trends. Nationally there developed a dominant migration pattern from East to West. After the opening of the border, and enforced by an economic crisis after unification, a dramatic depopulation of East Germany took place. Although the number of out-migrants has declined, there is still net migration from East to West. Since 1989, almost two million people left East Germany, a tenth of the population. Because of its demographic selectivity (it was especially those aged 20 to 35 years), this process had a dramatic effect on the demographic structure of the East German population, considered as even more significant than the Second World War (Bertram, 1996). A second effect was a dramatic drop in birth rates, similar to other Eastern European countries. Though it has recovered slightly during the last few years, it remains at a comparatively low level.

East German regions also faced a strong suburbanisation process, supplied by population flows from the cities and rural areas. Migration has mostly affected large cities and remoter rural areas, in particular in the north and in the eastern border regions. In these regions, the combination of suburbanisation and emigration to the West has led to a rapid ageing of the remaining population and a trend towards its feminisation.

In West Germany, mobility trends are more heterogeneous. The dominant trend is still migration from north to south but this is weakening. Mobility studies indicate that employment opportunity and rurality (low population density) are the most important positive factors explaining mobility. Consequently, some, but not all, rural areas show clear signs of counter-urbanisation processes. Yet, these trends are very selective. While there is an in-migration surplus in the population aged over 30 years, there is net loss among younger cohorts (Maretzke, 1998).

Economic Structures and (Un-)Employment

Although Germany is often identified with manufacturing, services are equally important for employment and are even more important for the national income. This is shown in Table 7.2. However, in an international comparison, Germany is still one of the more industrialised among modern economies. Unemployment levels are also comparatively high, a major source being the declining industries.

Table 7.2   Structure of the German economy in 1997

	Employment (%)	Gross domestic product (%)
Agriculture, forestry, etc.	2.7	1.3
Manufacturing and construction	31.3	30.8
Trade, transport and communications	24.3	17.1
Financial, renting and business	12.6	29.6
State and private services	29.1	21.3

Source: BMWi, 1999.

Following a demand-led boom after unification, initiated by the over-valuation of the East German Mark, the West German economy performed poorly. It was unable to compensate for the dramatic decline in the East German economy. During the 1990s the working population of Germany declined by two million, while the total population grew by 1.5 million, due largely to an immigration surplus. A sharp increase in unemployment levels could only be ameliorated by the massive use of early retirement measures, in particular in East Germany. Currently, while in West Germany unemployment is high, it is extremely high in East Germany.

Since the unemployment problem in West Germany is related to industrial decline, it is less surprising that rural areas' performance during the last two decades in generating jobs is much better than that of urban areas (Schrader, 1999). However, this performance, in particular in remoter areas, is to a large extent due to employment growth in manufacturing. Manufacturing still plays a significant role in rural areas. Services, in contrast, although contributing to employment growth for the whole economy, are still under-represented. Nonetheless, although we still find rural areas with structural problems, the generalisation of rural as backward and declining is certainly not true any more in West Germany. This is clear from Table 7.3.

Table 7.3   **Gross domestic product (GDP) per employed person and unemployment rate in rural areas of West and East Germany, 1997**

		Type I (agglomerations)	Type II (urbanised)	Type III (rural)	Overall
GDP (DM)	West	84,615	84,918	76,149	*99.313*
	East	53,106	54,614	57,188	*61.430*
Unemployment (%)	West	8.8	9.3	8.7	*10.4*
	East	17.3	19.9	20.5	*18.4*

Source: Schrader, 1999; data for rural Kreise as defined in footnote 1.

In East Germany we find a completely different picture, as can also be seen in Table 7.3. Economic performance is in general below any (rural and urban) West German region. Although disparities in productivity between rural and urban areas are not as high as in West Germany, unemployment rates in rural areas are higher. This is due to losses of employment opportunities in both manufacturing industries and agriculture (see below) after unification. Consequently, the whole of East Germany is an Objective 1 area and obtains priority funding through the European Union.

Agriculture

One of the most striking differences between East and West is in agriculture. Between 1949 and 1998, the number of farms in West Germany declined from about 1.6 million to less than half a million. In the same period the number of people employed in agriculture decreased from more than five million to about one million. Despite these significant changes, economists argue that agricultural restructuring did not lead to a significant change in the relative competitive position of West German agriculture. Indeed, despite huge changes over recent decades, agriculture in West Germany remains characterised as being dominated by small and medium sized family farms.

East Germany, by contrast, has had to deal with the heritage of socialistic collectivisation. Due to land reforms before 1949, the number of farms increased from about 450,000 to more than 700,000, before collectivisation and further steps towards integration led to a significant reduction in numbers. Drawing on official statistics, by 1989 a mere 8,668 farms existed in East

Table 7.4   Number of farms and farm sizes in Germany, 1998

		Farm size (ha)			
		0–50	50–200	200+	Total
West Germany	Number	417,941	64,192	1,558	*484,306*
	Share of farms	86%	13%	0%	*100%*
	Share of land	51%	45%	4%	*100%*
East Germany	Number	20,183	4,748	5,846	*31,997*
	Share of farms	66%	16%	18%	*100%*
	Share of land	4%	10%	86%	*100%*

Source: own calculation according to Agrarbericht (MB), 1999.

Germany, either as cooperatives or state farms (Wiegand, 1994). There were also 3,558 private farms, which were church owned or in individual ownership. They existed predominantly in horticulture and were very small (and officially neglected). A specific feature of East German agriculture, even among the former 'socialist' countries, was the separation of plant and animal production. Cooperatives and state farms in plant production used, on average, 4,100 hectares of land and cooperated with two to six farms specialised in animal production. The employment share of the agricultural sector was 9.6 per cent in 1989. Nonetheless, these statistics are not directly comparable to West European countries because agricultural firms in East Germany were highly diversified. They integrated agricultural services, non-agricultural production (in particular construction units) and a wide range of social and cultural activities. However, employment in the core agricultural activities was still much higher than in West Germany.

After unification, several partly competing political objectives existed in former East Germany, such as separating the business activities and social functions of agricultural firms, encouraging the (re-)establishment of family farming, dealing with severe environmental problems, increasing productivity and cushioning the dramatic social problems. Within three years, agricultural employment dropped from over 850,000 to 150,000 people. The dramatic effects of this loss could only be absorbed by widespread use of early retirement measures. Until 1993, it is estimated that about 180,000 people claimed early retirement (Schmidt and Neumetzler, 1993), which embraced almost the entire generation above 55 years of age. Despite the immense use of social policy measures, the dramatic decline of agricultural production, which went along with industrial decline, contributed significantly to the current high degree of unemployment in rural areas. Whilst older employees were absorbed by social security measures, and the younger generation out-migrated, it was middle aged women who were especially affected by unemployment in rural areas (Zierold et al., 1997). Current employment per hectare is even lower than in West Germany, due to the dramatic reduction of animal stocks and the lower capital intensity of production. Yet, despite the decline in employment, the number of agricultural units had grown to about 32,000 by 1998, as seen in Table 7.4. Although part-time family farms dominate in numbers, their contribution to employment and production is marginal. Instead of a transformation into Western style family farming, the outcome of agricultural restructuring in East Germany is a varied and novel mixture of capitalistic, cooperative and family business type farms. Common features of East German farms are large sizes in terms of land (Table 7.4) but low capital stocks, use of hired labour and almost one hundred per cent land tenancy. Economic pressures have led to an immense growth of productivity, but all farms still appear to be very vulnerable.

Policies for Rural Areas in the German Federal State

Regulation to Balance Regional Disparities

Germany is a federal state. Its constitution stresses the subsidiary principle and regions therefore are powerful. The various federal levels are outlined in Table 7.5.

Table 7.5   Administrative levels in the German federal system

	Administrative level	**Elected parliament**
Bundesrepublik	federal state	Bundestag
Bundesland	region	Landestag
Bezirksregierung (some Länder)		
Landkreis/Kreisfreie Stadt	county/town	Kreistag/Stadtrat
Gemeinde/Stadt	municipality	Gemeinderat
Dorf/Stadtbezirk	community, village	Ortsbeirat (some Länder), Bezirksvertretung (larger towns)

Note: levels with elected parliaments in italics.

Regional policy is both the right and the obligation of the regions. In principle, the federal state has only limited, indirect opportunities to influence regional policies. However, during the post-war period there has been a slight centralisation process in West Germany. This is legitimised by the objective of 'comparable living conditions' in the German constitution. A comprehensive regional policy framework was established in the 1960s and 1970s, which is still in place today and has been transferred to East Germany after unification.

The Bundesländer (regions) deliver all kinds of internal policies, from police and education up to economic development. The federal state's own institutions are limited to military, external affairs and border protection. There are some national bodies, but these are mostly formed as associations of regional organisations and the federal government. This construction limits the influence of the federal state on framing legislation or financial incentives. There are also policy arenas, such as education, where the federal state has no power at all, and coordination takes place in conferences of regional governments with autonomous power.

A further measure to overcome regional disparity is the Länderfinanz-ausgleich (regional financial adjustment). According to the federal structure, the financial constitution has to ensure that the different political bodies can fulfil their obligations. Tax revenues from the most important taxes (income tax, VAT) are shared between the federal state, the regions and the municipalities. Taxes are collected at the regional level and tax revenues are higher in more prosperous regions. The system of financial adjustment guarantees that poorer regions reach at least 95 per cent of the average per capita tax revenue, a constant bargaining issue among regional and national actors.

The federal state pays additional money for some regions to deal with specific budget problems for them to attain 99 per cent of the average per capita tax revenue. The latter recently prompted a review of the whole system by the Constitutional Court as it leads to a situation where some of the poorer regions end up with higher tax revenues than more prosperous regions. The Constitutional Court did not question the legitimacy of the system but imposed some reforms in favour of net contributors (BverfG, 1999). In recent years more than 40 billion DM have been redistributed annually, including about 25 billion DM paid by the federal government. Currently, only four of the 16 regions are net contributors. Finally, on the regional level, a similar system of financial adjustment exists to avoid large disparities between municipalities.

The Regional Planning System

In accordance with the federal structure, Germany has a hierarchical planning system (Henckel, 1999, pp. 251–72). The role of the federal state is to provide a legal framework (*Bundesraumordnungsgesetz*, BROG) and a framing national plan (*Bundesrahmenplan*). For rural areas, the maintenance of agricultural production, a sufficient population density, an adequate economic performance, the creation of jobs, landscape preservation, and the protection of nature are objectives formulated in the BROG, and concretised in the national plan. The legal framework is based on the principle of a proactive state. The legal framework and the national plan are further concretised and adapted to fit specific regional conditions by regional laws and development plans. The most important tools of the regional planning system are regional plans for specific areas to coordinate between the national, regional and local level. Within the framework of such plans, municipalities are responsible for land use in their territory.

Key concepts of the planning system include central places, settlement axis, functional divisions of territories and the definition of planning areas. The concept of central place has been the most important for the planning practice. It has led to a 'silent', passive restructuring in rural areas. The surplus of functions in central places went along with losses in the majority of smaller rural settlements. The definition of functional priorities for sub-regions has also been of some importance, in particular in connection with environmental protection. Defining the 'functions of rural areas' – areas of environmental compensation, recreation and agricultural production – is one of the immortal creations of the official language.

The Joint Tasks

Constitutional obligations of the federal state and the need for coordination of regional policies led to the establishment of 'joint tasks' (*Gemeinschaftsaufgaben*) between the federal state and the Bundesländer under West Germany's Great Coalition of the late 1960s. Two of these joint tasks are of particular importance for rural development:

- improvement of the farming structure (Gemeinschaftsaufgabe zur Verbesserung der Agrarstruktur und des Küstenschutzes, GAK); and
- improvement of the regional economic structure (Gemeinschaftsaufgabe zur Verbesserung der Regionalen Wirtschaftsstruktur, GRW).

The joint tasks are integrated into the legal framework of the constitution, which explains their overall stability during the last three decades. Yet, internally, objectives and programmes have undergone significant change.

The central body of each joint task is a planning committee (*Planungsausschuss*), constituted by representatives of the regional and the federal state. These committees have to agree on an annual framework plan (*Rahmenplan*) covering the principles of funding, programmes and initiatives. Regionally designed development programmes are co-financed on this basis. The federal state contributes 60 per cent of the budget (70 per cent in East German regions), which puts it in a strong position. There is a general concern about the legitimacy of the joint tasks (Johannes, 1998), as interweaving of executives means a restriction of parliamentary rights in the decision-making processes. The strong position of the federal government, due to its larger financial contribution, also raises the issue of this construction allowing the federal government to intervene in political issues, which are regional responsibilities. Certainly, this institution has changed the power structure and the position of local authorities.

The GAK is a second column of German agricultural policy. The first column is the agri-social policy, which includes measures and programmes such as social security for farmers or early retirement programmes to encourage farmers to leave the land. In 1998 the planned budget for all these measures was 2.7 billion DM. Most of this was directly dedicated to individual farms. Almost 60 per cent of these payments to farmers in West Germany are dedicated to naturally less favoured regions, which are not necessarily economically disadvantaged. Investment support is more important in East Germany because of an ongoing restructuring process, but will eventually decrease.

The GRW is the central instrument of regional development policies. Financial measures within the framework of the GRW have always been directed towards less favoured regions. A major problem was the definition of such regions. Although the framework itself has remained stable, objectives and indicators to identify less favoured regions have changed considerably during more than 25 years of existence (Struff, 1992, 1997). Since the late 1970s, indicators to measure unemployment, household incomes and infrastructure have been used to identify target regions. As a result of negotiations within a federal system, these target areas have always been comparatively large. In 1989, before unification, 48 per cent of West Germany, with more than a third of the population, was defined as areas to be supported under the GRW. Constant within the GRW, until 1994, was a special promotion for Kreise along the eastern border (*Zonenrandfoerderung*) and West Berlin.

The support of 'rural areas' has also been a permanent objective within the GRW. However, as Struff (1992, p. 165) argues, the old industrial regions always had a higher priority. Currently there are A, B and C zones. There are no distinctions between rural or urban areas. Most of East Germany is designated into A zones, with the highest priorities; some islands within East Germany, in particular surrounding Berlin, are defined as B zones; C zones are West German regions with disadvantages.

European Initiatives

Given the existence of an already well established system of regional development, European funding has not altered rural policy in Germany significantly, but has been integrated in the administrative routine of a well-established regional planning and economic development framework. However, financial effects have been quite significant, in particular in East Germany. There has been a mutual influence between European and national regional policies. While existing national programmes were accepted on the European level, national schemes were altered to allow co-financing of European initiatives. Alternatively, regions can co-finance initiatives on their own. This was necessary, in particular, for measures financed through the European Regional Development Fund in Objective 5b areas in West Germany, which only overlapped 50 per cent of the territory designated by the GRW. The fixing of the Objective 5b areas has been an internal process within the regional administration. Again experiences with such procedures already existed, because of the existence of the national programme. There has been little involvement at the local level, but huge negotiations between the regions, whose objective is to maximise external funds (see the section on local power and participation below).

It is no surprise that the measures financed within the Objective 5b framework are not new. The most popular individual programme of the 1994 to 1999 period was 'village renewal' (Tissen and Schrader, 1998), which has existed with alterations since the 1970s. A similar observation can be made about measures of rural and agro-tourism (Urlaub auf dem Bauernhof), which had a long tradition in development programmes in the south. Bavaria, which covers the largest share of the 5b area (42 per cent) and population (45 per cent), is the most important rural holiday resort area due to a policy of promoting pluriactivity on small farms in the post war period.

There is almost no comprehensive overview of the impacts of rural development schemes in East Germany, which is an Objective 1 area. The 'village renewal' programme has been even more popular in East Germany, due to the need for improvements in local infrastructure. In a period of budget constraints and cuts in public expenditure at the local level, village renewal has been the only scheme specifically designated for rural villages. Although it has an infrastructure focus, and is perceived as expert dominated, the refurbishment of village centres and symbolic buildings (churches, village halls, manor houses) has contributed positively to the local community. Nevertheless, the economic impacts are minor.

Context of Social Change

Changing Actors in Rural Areas in West Germany

Significant changes in social structures due to migration processes and changing occupation structures have shaped rural life in recent decades. In post-war West Germany, the influx of millions of refugees caused a significant population increase in rural areas. The refugees usually developed a parallel structure of clubs and informal networks to the old local establishment. Community studies during the early 1950s revealed that the assimilation of refugees was more difficult the more the local community was characterised by pure farming activities (Becker, 1997, p. 43). Although many of these new inhabitants out-migrated during the 1960s and 1970s, a significantly larger population remained as a lasting effect.

This historical legacy may be the reason that analysis of more recent migration into the countryside has not been identified as counterurbanisation processes as clearly as in other West European countries. Starting in the late 1970s, cheap rural housing, increasing mobility and a rising identification with nature and ecology promoted a rural population shift. Newcomers from the cities came to the villages. Young families were the key group (Becker, 1997), motivated usually by the prospect of the owning property in a pleasant rural setting.[2] Going along with the new migration trend, renting became more common in the countryside, too.

In the more accessible rural areas within driving distance to industrial towns, the new rural population is dominated by comparatively well paid employees commuting to these towns. In remoter rural areas, on the other hand, the new population consists of locally rather independent freelance workers, occasionally living in a second home in the countryside, and the 'alternative'[3] lifestyle oriented young. The latter live either on welfare or start projects in things such as eco-farming. There are also industrial employees in remote areas, working in subsidiaries of large companies that had been established with public subsidies under rural development policy. For some areas, retirement migration has also become significant. Finally, asylum seekers are also an important source of in-migration, being distributed equally over the country.

The migration processes are part of and go along with changes in the occupational structure of the countryside. Economically, the prevalence of farmers, local craftspeople, merchants and pub owners in the 1960s was diversified, for example, by trading and transport businesses, as structural adjustment provided less and less farm employment. Indeed, agricultural restructuring has had the most significant impact on the occupational structure. Before the vanishing peasant it was the agricultural worker which disappeared (Kromka, 1990). In the 1980s, even in remote rural areas, hardly more than 10 per cent of the work force was occupied in farming. Nonetheless, part-time farming and pluriactivity contributed to a relative stability. As long as young members of former farming families could somehow make their living in the villages their 'farming lifestyle' remained stable. The traditional agricultural lifestyle persisted as a dominant feature of rural life. However, more recent studies indicate a 're-professionalisation' of agriculture. This includes the prospect that only single farms will remain, while part-time farming will decline in importance (Becker, 1997, p. 156). In the place of agriculture, industry and, increasingly, the service sector must provide jobs.

There are at least two significant impacts of these trends on rural life. The first is the impact on the social composition of the rural population. Herrenknecht (1990) describes the changes on a community level as a shift from a single 'village culture' towards heterogeneous 'villages cultures'. He identifies four main groups:

- *old villagers*: indigenous people in the old centre of the village, circulating within the triangle of village hall, pub and church;
- *residential villagers*: a modernising culture living in the new housing estates, circulating within their own networks of private party rooms in people's houses and sport clubs;
- *emancipated villagers*: the Bildungsbürger or the 'alternative' movement, active in direct marketing, green policies and youth centres;
- *reripheral groups*: such as foreigners, asylum seekers and isolated newcomers.

The 'residential' and the 'emancipated' villagers challenge the indigenous establishment on the political level as well as on definition of 'legitimate tradition'.[4] A shift in the voting population in numerous villages and Landkreise saw new 'red-green' majorities take over in regions traditionally dominated by conservatives, sometimes abruptly changing basic policy patterns that they had fought against in the long opposition years. In the region of Luechow-Dannenberg, for example, a new red-green Landkreis coalition in 1994 stopped the acceptance of money granted to the region as compensation for a planned nuclear waste dump. Although within three years this led to a budget cut of roughly eight per cent and, since then, to significant budget deficits, this coalition gained continued support from a majority of voters in 1998.

A second impact is the expansion of individual actors' action space. Rural society has become an 'automobile society' (Laschewski, 1998c). The local economy has become integrated in rural economy networks (Becker, 1997, p. 104) of occupational opportunities as well as leisure time activities. Relevant politics are also located 'within the region' rather than confined to the local community. Herrenknecht (1990) has tried to express this with the concept of the regionalised village.

Social Change in East Germany

Post-1945, East Germany faced similar problems with the integration of refugees from former German territories. In many respects, integration in rural areas was more difficult due to greater destruction because of the war. One major objective of the 1945 land reform was to provide land for these refugees. With the reform, a new (heterogeneous) social group was created – the *Neubauern* (new peasants) (Bauernkämper, 1994). Due to differences in the historical farm structures, much more land was expropriated in the north than in the small peasant dominated rural regions of the south. This also explains why more refugees settled in the north. The *Neubauern* faced considerable difficulties due to lack of resources and, in many cases, skills, because they often did not come from rural regions. There were also conflicts between the 'old' and the 'new' peasants. Because of considerable economic problems, many *Neubauern* gave up rather quickly, either moving to West Germany or to work in the new industrial centres.

Collectivisation in agriculture was partly motivated by the problems of this new social group. In particular, in the early 1950s, new agricultural cooperatives founded by *Neubauern* struggled to survive (Laschewski, 1998a, pp. 32–3). The majority of the old peasants were forced into cooperatives but, in many cases, they did not join the existing cooperatives of the *Neubauern*. For almost a decade in many villages, cooperatives of old and new peasants existed alongside each other until forced to merge during the 1960s. However, recent studies showed that the distinctive role of the 'old' peasants has remained (Brauer et al., 1996; Laschewski, 1998a).

Unlike in the West, urbanisation as an East German migration trend continued until unification. Agricultural industrialisation and the allocation of manufacturing to rural towns determined intra-regional mobility. Due to the concentration of agricultural estates, a new hierarchy between villages emerged. Villages which housed the administrative centres of collective farms benefited, while smaller villages were neglected (Parade, 1991). Agricultural cooperatives and state owned farms provided social services and supported infrastructure building. Chairs of cooperatives were more influential than local mayors. Due to the monopoly of the Socialist Unity Party and the centralisation of political decisions, the role of municipalities was limited to mobilising the population for 'voluntary' actions and mass political events, such as May Day celebrations (Siebert and Laschewski, 1999).

The specific process of agricultural industrialisation under East German socialism had significant effects on local power structures. In theory, a dual hierarchy of the agricultural technocratic elite – the cadres of the cooperatives – and the party elite replaced the old rural elite. Mechanisms were put in place in the 1960s and 1970s to weaken the old elite, such as restrictions on the children of landowners accessing universities. However, although formally disadvantaged, the old elite often remained a strong local force because of informal relationships. This became apparent after unification in the agricultural privatisation process (Laschewski, 1998a). Informal relationships were of key importance because of the permanent shortage of goods. The planned economy produced an extensive informal exchange economy as a parallel world. Yet, in particular in the north, in some regions a true peasantry never existed, and in other regions it ceased to exist because of exile during the 1950s. In such communities, the technocratic elite, usually recruited from the south, stepped into a position similar to former farm 'inspectors' (management), while the Party replaced the 'paternalistic' absent, feudal landowner. In these regions, agricultural employees remained what they had always been – farm workers.

Of particular interest are gender relations. Although women were usually formally fully employed, a gendered division of labour was not eliminated. Women's employment was concentrated in the social services, administration and animal production. In agricultural production, they tended to do less manual and more repetitive work, and were likely to be at a lower level in the hierarchy. Women usually remained responsible for all household work, despite being in paid employment.

Unification brought several significant changes to this post 1945 rural structure. These included the following:

- agricultural cooperatives lost their almost monopolistic position in providing jobs, as well as their function to provide social ser.ices to the community;
- women were affected strongly, and are often still unemployed in rural;
- the Socialist Unity Party was abolished, and lost its status as the 'legitimate' monopoly on public policy issues;
- landowners were 'passive' winners, due to the reinstallation of property rights;
- informal networks lost their economic function due to a better supply of goods, the successful establishment of social security systems and economic restructuring. Migration and long-distance commuting also affected people's ability to participate in local activities;
- there was an influx of West Germans, usually in higher occupational positions in the private sector (chartered accountants, lawyers, consultants), but also in the public sector and, occasionally, in agriculture.

The primary dividing line of winners and losers in former East Germany is between having a job and being unemployed. As a general rule, the higher the position in the occupational hierarchy the greater was the chance of staying employed. Those who were employed in sectors which were privatised were more likely to lose their jobs than those who were taken over by public services. There are also several jobs, almost inaccessible for East Germans, which were usually taken by West Germans. Replacement by West Germans in higher positions in the public services forced former cadres (including university professors) to become entrepreneurs or functionaries in the intermediary sector.

A second dividing line is along the distribution of property rights. If locally in existence, landowners were usually in a strong position to re-establish their domination in agriculture, although this did not necessarily mean a return to family farming (Laschewski, 1998a, 1998b). Elsewhere, it was either cadres of the former cooperatives or West Germans who tried to establish farms. However, apart from very remote rural areas, there was also a power shift from agricultural to non-agricultural activities. In many communities, new entrepreneurs (often former cadres) and public servants in higher positions (often West German) have come to form a new local elite.

Problems of unemployment have a dramatic gender dimension in rural areas. For example, the unemployment rate in some rural Kreise in Brandenburg is almost 25 per cent, when the average for the state is 15.1 per cent. Yet, evidence for individual villages shows an unemployment rate of up to 90 per cent for women (Koeppl, 1997). While men might find a job by migration or by driving longer distances to their workplace, the unemployment rate for women in the countryside remains stable and presents a major obstacle to improving the life conditions of women in the East. On this point, the difference between East and West might be most significant. The old production cooperative was more than just a place for earning one's money, it was a place for interaction, communication and social embeddedness (De Soto and Panzig, 1995; Rocksloh-Papendieck, 1995).

The Social Representation of Rurality

In Germany, 'rurality' is a secondary concept, usually subordinated to more dominant ideas such as 'region', 'peasant' or 'periphery'/'border'. The political administrative representation of rurality can be discovered in regionality as it is this rather than rurality that provides the principal orientation. In Germany, as noted before, the regional administrative structure, as well as the historical construction of autonomous Länder, have built strong local power. Reflecting this status, within official statistics 'rural' does not provide a primary category. Unlike other countries, such as Britain, communities are not defined as urban or rural. There is only one secondary statistical categorisation of rurality, which we described earlier, provided by a federal institute for regional planning. Rurality in this sense is a regional characterisation among others.

The representation of rurality within regional policy is twofold. First, rurality is connected with economic 'backwardness'. This rather classical perception has recently been challenged by the good recent economic performance of many rural regions. However, Struff (1997, pp. 91–2) sees its persistence among regional economists and planners, where the supposed disadvantages of rural regions are deduced from particular advantages of agglomerations. Second, rurality is closely linked to the (national) periphery. Before unification, the 'iron curtain' was a huge problem for rural development on both sides of the border. In West Germany, the predominantly rural areas along the East German border received special grants within the GRW, as noted above. Unification changed this geography, with these formerly peripheral parts of East and West Germany suddenly situated right in the centre.

Alongside the regional planning system, rurality is linked to ideas of the 'peasant' and the 'village'. Both concepts have also been at the core of rural sociology, which is more 'agricultural sociology' (*Agrarsoziologie*) than 'rural sociology'.[5] Thus, both peasants and the village have been discussed under the general idea of modernisation (Kromka, 1990). Most empirical work has been done either on peasant mentalities and households (Pongratz, 1990) or as community (village) studies (Struff, 2000). Rurality, therefore, has been constructed as pre-modern, either positively as a resort of cultural distinctiveness, or negatively with a notion of a 'cultural lag'. There have been sharp attacks by 'modernisers' that their opponents' position is a 'peasant ideology' (Bauerntumsideologie), alluding also to Nazi ideology. On the other hand, small family farms have served as symbol of 'free' capitalism and against socialistic suppression. Here we also find a connection to the regional policies in the Zonenrandgebiete,[6] the West German border region to the East. These concepts of 'peasant' and 'village' underlay West German agricultural policies, which are in principle still in place. The contradictory notions of the peasant help explain contradictions within these policies, which include protectionist as well as modernising elements. It is also noteworthy that the 'village renewal' programme is the only non-agricultural rural development programme under the agricultural policy framework.

The environmental or 'alternative' movement transformed the conflict between 'peasant ideologists' and 'modernisers'. The ecological debate, combined with the idea of 'back to nature', was an important aspect in the discourse of rural–urban dichotomy by the end of the twentieth century. A revival of the old dualism found rurality in a position to play the opposite part of industry, with its dirty damaging image. Agriculture and rural life were supposed to offer a safe and environmentally sound development, different to that of a condemned industrial system located primarily in the polluted urban areas. The green movement in Germany started with the discourse on *Waldsterben* (forest death) and immediately critiqued the industrial system. Problems with the sources of emissions and of air pollution were the first topics to become symbols of the alternative movement. This was accompanied by a search for new ways of life, where rurality and rural life were set up in opposition to a negative image of urbanity.

The countryside (*Das Land*) has thus gained a new meaning from that of a space subordinated by the urban region, in principal a Marxist approach. It is again peripheral, but not in a national sense. Instead, it has assumed a normative status (Bodenstedt, 1990), opposed both to the industrialised-urban way of life and to 'conventional' farming. Rurality, peasants and village life need protection against the destructive forces of modern capitalism. Unsurprisingly, this approach has been heavily attacked from the remaining 'modernisers', who also try to link this new 'eco-socialism' with Nazi ideas (Kromka, 1992). Yet, despite the fact that some ideas of this movement are supportive of projects which are not accepted by indigenous rural people, such as building houses with soil or starting-up new (eco-) farms (Höger, 1996), the ideas of the green movement have been institutionalised in regional political programmes. In Hesse, the environmental movement successfully established a development programme based on ideas of self-reliance and endogenous rural development when the Green Party became part of the regional government. LEADER's application in West Germany saw a further step in a process in which 'alternative' ideas became mainstream and integrated into the established framework of rural development (Bruckmeier, 2000). Within this process, rurality is transformed from a political concept into a marketing idea (Höger, 1996).

Referring back to the earlier classification of social groups by Herrenknecht (1990), we find that different representations of rurality can be related to different groups. However, as he indicates, local people, in particular the younger generation, tend to be familiar with several identities, and like the shift between different cultures. Cultures become consumables. This concurrency and a degree of equivalence between representations of the rural is also reflected in the current political context, in which new ideas of rurality only gradually transform the existing political framework.[7]

Local Power and Participation

The Retreat from Local Democracy

Local power relations in rural areas have undergone significant changes during recent decades, favouring the regions in particular. Vertically, with the evolution of regional planning and the development system, local government has become more and more subordinated to decision-making processes and regulations at higher levels of the institutional hierarchy. The relatively high communal and local autonomy of former times has been weakened. Also, due to the concentration of political power and economic resources at politically higher levels, more bureaucracy now features in communal institutions.

The political system in Germany is characterised by regulation through law and order. This influences decision-making processes in agriculture and rural development policies. Regional policy in Germany presents itself as a fairly comprehensive, administration-dominated (bureaucratic) system. Central government is limited to the use of legal means or financial incentives – a highly formalised planning system and the redistribution of money. Although there have been precise goals behind initial policy, continuous renegotiation between the regions has always led to a final integrative policy of satisfying everybody. For instance, the designated areas within the GRW have always covered about 50 per cent of the territory and more than a third of the population (Struff, 1997). The designation of rural development areas in regions such as Hesse and Bavaria, which are net contributors within the regional financial adjustment system, appears as a negotiated compromise to get some money back or to obtain a bigger share of the cake.

Bargaining amongst the Länder and with the federal state in former West Germany is thus a characteristic feature of regional policy. However, the system is about to be changed according to the draft of EU's Agenda 2000. The subsidy flow will change its course, becoming independent from negotiations between the regions and central government, and the main issues of negotiation will revolve around EU policy (Eltges, 1998). Yet, because of huge differences in economic performance, Agenda 2000 is going to reproduce the East-West divide. Rural development policies will, apart from a small area at the former German-German border, only take place in East Germany.

There is a strong domination by civil servants of national as well as regional ministries within the processes of political decision-making. The construction of the joint tasks by the federal and regional governments increased this problem. Within these negotiations, regions are not represented by parliaments but by the ministries and government departments. This lack of parliamentary involvement has been heavily criticised (Johannes, 1998). On the other hand, organised interests, such as farmers' unions, do have a considerable influence on decision-making processes.

Centralisation and bureaucratisation have undermined local autonomy in the West and are definitely penetrating the 'rural life world' (everyday life). Municipalities are more and more regulated and, therefore, limited in their options. Their budget is increasingly dependent on appropriated money. They have been subject to communal reforms which abolished the smaller local parliaments at the village level (Henckel, 1999). In practice, little remains of local self-administration other than the competence to designate housing areas. Again, differences between East and West, as well as regional specifications, need to be understood as influential for regional autonomy and the self regulating capacities of communal life. For example, limitations derive from unemployment levels, because communities are responsible for the delivery of social benefits for the poorest. This leads to the peculiar situation that those communities who need investments most have to spend comparatively more on private consumption.

Regulating Rural Conflicts

While local democracy has been sacrificed to achieve better administrative efficiency, the conflict regulation function of local councils has also been lost. More and more, extra-parliamentary forms of protest have become an important element within rural conflict regulation. The effectiveness of protest depends on the ability to organise political demonstrations and legal challenges. Therefore, people's participation becomes an important, newly induced issue. It is within this context that the environmental movement in West Germany created the concept of autonomous, regional development a decade before LEADER (Bruckmeier, 2000; Pongratz and Kreil, 1991). This may explain the doubts within official regional policy about the establishment of LEADER.

Within both LEADER programmes it was state administration at the Kreis level which dominated the establishment of initiatives and local action groups (if implemented at all) (Geißendoerferm et al., 1998; Tissen, 1998). LEADER I, in particular, was perceived as an unwanted interference from the European level in regional affairs and found little support, although there was a positive response at the local policy level. Subsequently, it seems that LEADER has become incorporated into the existing structure of local associations, unions and church organisations, which were already entrenched within policy making. As Bruckmeier (2000, p. 221) observes: 'More and more of the traditional economic groups and institutions returned to the policy process as beneficiaries of LEADER project funding, re-entering endogenous rural development on a road originally paved for other purposes and actors.' Such an eventual interweaving of corporate actors is typical of the functioning of the German political system, where organised interests dominate decision-making processes. Independent initiatives are rare because of the high degree of bureaucracy in the application process and the lack of co-financing, but also because many of the objectives are part of the duties of local authorities anyway. This seems to question the legitimacy of supposedly public initiatives within the given institutional structure of Germany.

There are other examples. The leading actors here are in many cases the 'alternative' new rural population. The remnants of the environmental movement today are among the principal driving forces for a new rural identity and for integrated and participative development approaches. To make a living in the countryside, these new rural people had to start projects, and hardly any project was started without some kind of public subsidy. Subsidies were granted for job creation, cultural or social initiatives, environmentally friendly agricultural production, or for direct or local marketing. In addition, subsidies were granted for especially innovative projects, partly in cooperation with universities, to explore new and alternative ways of rural development. However, these subsidies were commonly not intended to form part of an 'integrated' strategy. In fact, in many instances they prompted disintegration by providing external support for local residents who were questioning the local order of things. Eco-farming, communal living experiences and the sheer appearance of nonconformist nonresident young people threatened the local institutional structure described above. Villagers, in most cases, did not vigorously defend their local order and institutionalised habits, as they might have done a hundred years ago. This surely was due partly to the fact that disintegration of their community had already begun, but also because the new rural population had bought old farmhouses, rented rooms in other houses and demanded local construction services – they had bribed the old lifestyle with money. Local residents often tried to stop project growth politically, for example, by not granting construction permits via the municipal council. LEADER in West Germany can be seen as a further step of a process in which 'alternative' ideas became mainstream and are integrated in the established framework of rural development (Bruckmeier, 2000; Höger, 1996).

In summary, the political programme of 'people's participation' is a strategy or at least a reaction against the negative side effects of earlier structural changes and political decisions in the direction of centralisation and centralisation. However, within the existing political framework it favours social groups which are able to establish connections to national and regional power centres. Many participative rural development initiatives therefore establish parallel institutions alongside existing local authorities and the existing agriculture policy network, and so fail to be 'integrative' at all.

Participation in the Former East Germany

In East Germany the situation is different. The 'symbiotic relationship' between community and the former agricultural cooperatives (Herrenknecht, 1995) has been abolished. The political and economic spheres have been separated. Therefore, the municipalities have gained more power, which is executed by a variety of new and old actors. Local communities' dependence on agriculturally based establishments has been replaced by the dependence on external funding. This state domination also exists in the voluntary sector. This is all perhaps not surprising considering the socialistic past. The latter also helps explain the lack of social movements which could play a similar role to the 'alternative' movement in West Germany. Any evolution of strong voluntary institutions as the foundation for a new civil society also face the huge obstacles of the economic crisis, unemployment, radical social change and out-migration. Under such conditions, long term perspectives of participatory approaches are challenged by quick 'top-down' solutions. Investments in physical infrastructure have had a high priority in the unification process, which has strengthened even more the position of external (national or regional) actors. However, the obvious failure of such an approach to establish sustainable structures may eventually increase the chance for local participation.

Currently, participatory approaches face huge difficulties in the former East Germany. LEADER, for instance, appears as a tool to facilitate rather conventional ideas of rural development (Bruckmeier, 2000). It is very likely that the initiators of LEADER initiatives in East Germany are the same elites who are the driving forces behind most other activities. On the one hand, these are the Kreis administration and the organised political networks. On the other hand, they are the former East German functionaries, qualified staff of former cooperatives, academics, and so on, who are now engaging in new projects and businesses.

Lessons may also be learned from attempts to integrate women into local initiatives. As mentioned earlier, women are significantly disadvantaged within local labour markets. One important East–West difference is the attitude towards unpaid voluntary work. East German women are resisting the strong tendency to push women into voluntary 'jobs' and social activities and show little interest in this type of 'uneconomic' engagement. Recognising that the new cultural standards of money and economic calculation are among the reasons for their unemployment and misery, 'Many women in rural areas stubbornly insist that they be included in paid work situations. They refuse to commit themselves to participation in other activities as long as this condition is not fulfilled' (Koeppl, 1997, p. 22). Their rejection is well justified but especially unfortunate because 'civic and community work represents a major opportunity, besides regular employment, for gaining self-respect and the recognition of others' (ibid.). A similar response is given to the recommendation to women in the East to take part in the activities concerning Local Agenda 21 (Iganski, 1999). Those who have a background other than home-based activities reject such 'therapy'. They might be willing to work, but not without payment. To find a realistic course of action requires a more people-oriented approach (Teherani-Krönner, 1997).

Women in East Germany are proud to have managed their multiple burden in the past, when they felt 'emancipated'. They fondly remember the social context of their workplace, including numerous festivities. Unification truncated socialist policies designed to enhance women's role (Doelling, 1991; Nickel, 1992; Shaffer, 1981) and there is high consensus within women's studies that women have lost what they had perceived as beneficial policies in the previous political system. Moreover, women are now frequently discriminated against, especially when single and with children (Nickel, 1992; De Soto and Panzig, 1995). To a number of women, the discontinuity of employment and the collective has become equivalent to the loss of quality of life.

Iganski (1999) notes a research finding that as women's qualifications and levels of education rise they become involved in voluntary work to a greater extent. It seems justifiable to argue that the reverse may be true in former East Germany; as the level of professional involvement decreases the willingness for voluntary work also declines. This is a dilemma for policy-makers who prioritise direct resources for structural improvements within a 'central-town' system and then have to depend largely on voluntary initiatives to revive social structures in rural areas (Iganski, 1999).

Conclusion

Germany is a heterogeneous country, which has always been characterised by a strong regional autonomy, a characteristic still represented in the federal political system. Beside having a comparatively homogenous settlement structure, we find some urban agglomerations as well. Rural areas in Germany are, in comparison to other European countries, densely populated. The separation into East and a West Germany during the post-1945 period accounts for huge differences in development standards and structures between East and West, which overlay all other differences between German regions.

After unification in 1990 the political system, including the regional planning system which evolved in West Germany after the Second World War, was transferred to cover the East, too. Basic characteristics of this system are the high degree of framing regulation and legislation, and a complex negotiation system between regions and the federal government in the framework of the joint tasks. Within these development processes, more and more power and functions have been transferred from the municipalities to the regions and central government. It is within this context that changes in rural demography and in the occupational structure of the rural economy have brought about new demands for participation, in order to overcome varied forms of marginalisation.

In summary, these structural and institutional conditions help to explain why regionality appears to be more important than any discourse on rurality for understanding the dynamics of leadership, local power and development in Germany today.

Notes

1 Official regional statistics and most regional studies in Germany refer to the regional typology proposed by the Bundesamt für Bauwesen und Raumordnung (Federal Office for Construction and Regional Planning). This typology uses the distance to the next central town (Oberzentrum) and the population density on the administrative level of a Kreis. Within this typology three different types of regions and for each region four types of Kreise are defined. However, because some sub-categories do not apply in all regions, nine types of Kreise are defined. Three of them are defined as rural:
   • in regions with large agglomerations (Type I);
   • in urbanized regions with agglomeration tendencies (Type II); and
   • in rural regions (Type III).
2 Renting is much more common than living in one's own property in Germany. This is different in the countryside, where living in one's own house is common (Struff, 1992).
3 In Germany the environmental movement is characterised as 'alternative' movement.
4 See the next section on the issue of 'rural culture'.
5 Indeed, Barlösius (1995) argues that, with a declining numbers of farms and general modernisation, German rural sociology is in danger of losing its research subject.
6 The term itself indicates the political relevance. It refers to the situation, when Germany was separated into four occupied 'zones' by the allies after 1945. The existence German Democratic Republic was not acknowledged by the Federal Republic. It remained the East or Soviet Occupied Zone (Sowjetisch besetzte Zone). Consequently, the border regions were called 'zone border regions'.
7 This observation is similar to the argument that a movement towards a (literally) post-productivist rural society is easier to locate on the rhetorical than on the structural and institutional levels of society (Tovey, 1998).

References

Agrarbericht (various years), *Agrar - und ernährungspolitischer Bericht der Bundesregierung*, Bundesministerium für Ernährung, Landwirtschaft und Forsten, Bonn.
Barlösius, E. (1995), 'Worüber Forscht die Deutsche Agrarsoziologie?', *Kölner Zeitschrift für Soziologie und Sozialpsychologie*, vol. 47, pp. 319–38.

Bauernkämper, A. (1994), 'Von der Bodenreform zur Kollektivierung – Zum Wandel der ländlichen Gesellschaft in der sowjetischen Besatzungszone Deutschlands und der DDR 1945–1952', in H. Kälbe, J. Kocka and H. Zwahr (eds), *Sozialgeschichte der DDR*, Klett Cotta, Stuttgart, pp. 119–43.

Becker, H. (1997), *Dörfer Heute – Ländliche Lebensverhältnisse im Wandel 1952, 1972 und 1993/1995*, Schriftenreihe der Forschungsgesellschaft für Agrarpolitik und Agrarsoziologie e.V., Bonn, No. 307.

Bertram, H. (1996), 'Familienstrukturen und Haushaltsstrukturen', in W. Strubelt, J. Genosko, H. Bertram, J. Friedrichs, P. Gans, H. Häußermann, U. Herlyn and H. Sahner (eds), *Städte und Regionen – Räumliche Folgen des Transformationsprozesses*, Leske and Buderich, Opladen, pp. 184–287.

BflR (Bundesforschungsanstalt für Landeskunde und Raumordnung) (1995), Laufende Raumbeobachtung, *Materialien zur Raumentwicklung*, 67.

BMELF (Bundesministerium für Ernährung, Landwirtschaft und Forsten) (1999), *Die Verbesserung der Agrarstruktur in der Bundesrepublik Deutschland 1994–1996*, BMELF, Bonn.

BMWi (Bundesministerium für Wirtschaft und Technologie) (1999), *Wirtschaft in Zahlen 1999*, BMWi, Berlin.

Bodenstedt, A. (1990), 'Rural Culture – A New concept', *Sociologia Ruralis*, vol. 30, pp. 5–47.

Brauer, K., Willisch, A. and Ernst, F. (1996), 'Intergenerationelle Beziehungen, Lebenslaufperspektiven und Familie im Spannungsfeld von Kollektivierung und Transformation', in L. Clausen (ed), *Gesellschaften im Umbruch – Verhandlungen des 27. Kongresses der Deutschen Gesellschaft für Soziologie in Halle an der Saale 1995*, Campus, Frankfurt a.M./New York, pp. 736–49.

Bruckmeier, K. (2000), 'LEADER in Germany and the Discourse of Autonomous Regional Development', *Sociologia Ruralis*, vol. 40, pp. 219–27.

BverfG (Bundesverfassungsgericht) (1999), Urteil des BVerfG, 2 BvF 2/98 vom 11.11.1999, Absatz-Nr. (1–347), http://www.bverfg.de/.

De Soto, H. and Panzig, C. (1995), 'Women, Gender and Rural Development', in U. Altmann and P. Teherani-Krönner (eds), *Frauen in der Ländlichen Entwicklung. I. Fachtagung*, HUB, Berlin, pp. 111–30.

Doelling, I. (1991), 'Between Hope and Helplessness: Women in the GDR After the Turning Point', *Feminist Review*, vol. 39, pp. 3–15.

Eltges, M. (1998), 'AGENDA 2000: was Bedeuten die Strukturpolitischen Vorschläge für die Bundesrepublik Deutschland?', in *Informationen zur Raumentwicklung*, BBR, Bonn, pp. 567–78.

Geissendörfer, M., Seibert, O. and von Meyer, H. (1998), 'Ex Post-evaluierung der Gemeinschaftsinitiative LEADER I in Deutschland', *Berichte über Landwirtschaft*, vol. 76, pp. 540–79.

Henkel, G. (1999), *Der Ländliche Raum*, 3rd edition, Teubner B.G., Stuttgart.

Herrenknecht, A. (1990), 'Das Dorf in der Region – oder: Steht die Dorfdiskussion vor einem Paradigmen-Wechsel?', *Pro Regio*, vol. 12, pp. 13–9.

Herrenknecht, A. (1995), 'Das Dorf in der DDR, Dorfbilder, Dorfdisskussion und Dorfentwicklung von 1960–1989', *Pro Regio*, vol. 17.

Höger, U. (1996), 'Von der Eigenständigen zur Eta(t)blierten Regional-entwicklung', *Pro Regio*, vol. 18–19, pp. 15–8.

Iganski, B. (1999), 'From "Superwomen" to "Prosperity Trash": Women in Rural Development in the former GDR', in P. Teherani-Krönner, U. Hoffmann-Altmann and U. Schultz (eds), *Frauen und Nachhaltige lä endliche Entwicklung*, Centaurus, Pfaffenweiler, pp. 173–9.

Johannes, F. (1998), 'Finanzverfassungspolitische Aspekte der Gemeinschafts-aufgabe', in BMELF (Bundesministerium für Ernährung, Landwirtschaft und Forsten) (ed.), *25 Rahmenplan der Gemeinschaftsaufgabe – Verbesserung der Agrarstruktur und des Küstenschutzes*, BMELF, Bonn, pp. 41–51.

Koeppel, U. (1997), 'Promotion of women in rural projects of Brandenburg', in P. Teherani-Krönner and U. Altmann (eds), *What Have Women's Projects Accomplished So Far? II. International Conference: Women in Rural Development*, HUB, Berlin, pp. 17–29.

Kontuly, T. and Dearden, B. (1998), 'Regionale Umverteilungsprozesse der Bevölkerung in Europa seit 1970', in *Informationen zur Raumentwicklung*, BBR, Bonn, pp. 713–22.

Kromka, F. (1990), 'Vier Jahrzehnte Westdeutsche Land- und Agrarsoziologie', in G. Vonderach (ed.), *Sozialforschung und Ländliche Lebensweisen*, WVB, Bamberg, pp. 3–31.

Kromka, F. (1992), 'Zwischen Agrarromantik und Agrarfeindschaft: Die Irrtümer der ökosozialistischen Agrarsoziologie', *Agrarwirtschaft*, vol. 41, pp. 280–89.

Kroner, G. (1993), *Konzeptionelle Ansätze zur Abgrenzung Ländlicher Räume*, Working Paper, Akademie für Raumforschung und Landesplanung, Hannover.

Laschewski, L. (1998a), *Von der LPG zur Agrargenossenschaft – Untersuchungen zur Transformation Genossenschaftlichen Organisierter Agrarunternehmen in Ostdeutschland*, Edition Sigma, Berlin.

Laschewski, L. (1998b), 'Continuity and Changes: Agricultural Restructuring in East Germany', *Eastern European Countryside*, vol. 4, pp. 37–48.

Laschewski, L. (1998c), 'Risikogesellschaft auf dem Lande als Risikogemeinschaft?', *Kirche im Ländlichen Raum*, pp. 139–43.

Maretzke, St (1998), 'Regionale Wanderungsprozesse in Deutschland sechs Jahre nach der Vereinigung', in *Informationen zur Raumentwicklung*, BBR, Bonn, pp. 743–62.

Nickel, H. (1992), 'Geschlechterbeziehung und – Sozialisation in der Wende. Modernisierungsschübe oder – Brüche?', *Berliner Journal für Soziologie*, vol. 3, pp. 381–7.

Parade, L. (1991), *Das Leben in den Dörfern: vor, Während und Nach der Wende*, paper presented to the Second European Village Renewal Congress, Reichenbach.

Pongratz, H. (1990), 'Cultural Tradition and Social Change in Agriculture', *Sociologia Ruralis*, vol. 30, pp. 5–12.

Pongratz, H. and Kreil, M. (1991), 'Möglichkeiten einer Eigenständigen Regionalentwicklung', *Zeitschrift für Agargeschichte und Agrarsoziologie*, vol. 39, pp. 91–111.

Rocksloh-Papendieck, B. (1995), *Verlust der Kollektiven Bindung. Frauenalltag in der Wende*, Centaurus, Pfaffenweiler.

Schaffer, H. (1981), *Women in the Two Germanies. A Comparative Study of a Socialist and a Non-Socialist Society*, Oxford University Press, Oxford.

Schmidt, K. and Neumetzler, H. (1993), *Strukturelle Anpassungsprozeß der Ostdeutschen Landwirtschaft im Blickpunkt von Beschäftigung und Erwerbstätigkeit*, Institut für Wirtschaftsforschung, Berlin/Halle.

Schrader, H. (1999), 'Tendenzen und Perspektiven der Entwicklung ländlicher Räume', in P. Mehl (ed.), *Agrarstruktur und Ländliche Räume: Rückblick und Ausblick*, Landbauforschung Völkenrode, Braunschweig, pp. 213–39.

Siebert, R. and Laschewski, L. (1999), *Becoming Part of the Union: Changing Rurality in East Germany*, paper presented to the XVIII European Congress for Rural Sociology, Lund.

Struff, R. (1992), *Regionale Lebensverhältnisse, Teil 1: Wohnen, Arbeiten und Sozialhilfe*, Forschungsgesellschaft für Agrarpolitik und Agrarsoziologie, Bonn.

Struff, R. (1997), *Entwicklung Ländlicher Räume und Regionale Wirtschaftspolitik in Deutschland*, working paper, Forschungsgesellschaft für Agrarpolitik und Agrarsoziologie, Bonn.

Struff, R. (2000), *Regionale Lebensverhältnisse, Teil 2: Sozialwissenschaftliche Dorf- und Gemeindestudien*, Forschungsgesellschaft für Agrarpolitik und Agrarsoziologie, Bonn.

Teherani-Krönner, P. (1997), 'Veraenderung von Handlungsspielraeumen von Frauen in Agrarkulturen', in D. Steiner (ed.), *Mensch und Lebensraum. Fragen zu Identität und Wissen*, Westdeutscher Verlag, Oplanden, pp. 267–89.

Tissen, G. (1998), *Zwischenbewertung der Gemeinschaftsinitiative LEADER II in Nordrhein-Westfalen*, FAL Institut fuer Strukturforschung, Braunschweig.

Tissen, G. and Schrader, H. (1998), *Foerderung Ländlicher Entwicklung durch die Europäischen Strukturfonds in Deutschland*, FAL Institut fuer Strukturforschung, Braunschweig.

Tovey, H. (1998), 'Rural Actors, Food and the Post-modern Transition', in L. Granberg and I. Kovách (eds), *Actors on the Changing European Countryside*, Institute for Political Science, Budapest, pp. 20–43.

Wiegand, S. (1994), *Landwirtschaft in den Neuen Bundesländern: Struktur, Probleme und zukünftige Entwicklung*, Vauk, Kiel.

Zierold, K., Laschewski, L. and Dippmann, L. (1997), 'Landwirtschaft im Ländlichen Raum; Funktionen, Formen, Konflikte: Zwei Vergleichende Explorative Fallstudien in Ostdeutschland', in Landwirtschaftliche Rentenbank (ed.), *Landwirtschaft im Ländlichen Raum – Funktionen, Formen, Konflikte*, Landwirtschaftliche Rentenbank, Frankfurt a.M., pp. 53–91.

Chapter Eight

Rural Restructuring and the Effects of Rural Development Policies in Spain

Fernando Garrido, José R. Mauleón and Eduardo Moyano

Introduction

Over the last 20 years, the experience of Spanish rural areas has been transformed. During the 1960s and 1970s, rural Spain was characterised by an exodus brought about by agricultural modernisation and the growth of the industrial sector in non-rural areas. Over the last two decades, in contrast, a new rural restructuring has taken place. This is marked by a combination of endogenous and exogenous factors: political (the democratisation process, political decentralisation to the regions, the entry of Spain into the European Union), cultural (the expansion of new post-materialist values emphasising the local level and the preservation of environment), and socioeconomic (the emergence of new non-farming actors in rural areas) (Pérez Yruela, 1995; Giménez Guerrero, 1998; Moyano, 1999). Such a restructuring has also been influenced by rural development policies.

This chapter analyses this transformation in five sections. The first describes the administrative institutions existing in Spain and the jurisdiction of each of them in relation to policies for rural development. The second section attempts to quantify the multiple forms followed by rural restructuring during the last decades in each of the Spanish regions. The third section describes two of the main rural development programs implemented in Spain: LEADER and PRODER. The fourth section makes a detailed analysis of the effects of both programmes in one of the principal rural regions, Andalusia. The final section brings these threads together to raise the issue of whether or not rurality is becoming redefined in Spain in the light of this restructuring process.

Administrative Structures in Spain

To understand both socioeconomic data on rural Spain and rural development practices it is necessary to outline first the administrative framework. Spain has a decentralised system of government administration organised into four tiers: national, regional, provincial and local, each with its own areas of responsibility.

At the national level, the government exercises power through a central administrative structure in Madrid, and through offices in the capital cities of the 50 provinces into which Spain is divided. However, these powers are being increasingly reduced, as responsibilities are transferred to the 17 regional governments, shown in Figure 8.1, established within the political decentralisation process of the 1978 Constitution.

Figure 8.1   Regional governments in Spain

Source: *Memoria Anual 1998*, Ministerio de Administraciones Públicas, Madrid, 1998.

As a consequence of decentralisation, the 17 regional governments are accumulating powers through the transfer of both responsibilities and economic and human resources from central government. The resulting model of the state cannot, in a strictly theoretical sense, be termed federal, since it is constructed from the top down, through the transfer of responsibilities from central government to the regions without any recognition of the right to self-determination. In practice, however, the levels of self-government and the number of powers assumed by each regional government are comparable with those of the German *länder*. At each region, these powers are exercised through an executive (the regional government), a legislative (the regionally elected parliament with the right to dictate its own laws), and a judiciary (represented by the law courts, which have the power to decide on issues related to the application of regional laws).

Each of Spain's 50 provinces has a government (Diputación). These bring together the local municipios in each province, and are financed from the resources assigned by central government to local authorities according to the size of their population. Provincial governments have the responsibility, delegated up to them by their local authorities (Ayuntamientos), for reducing territorial inequalities in the services provided to their citizens. Their powers, nonetheless, except in the Basque Country and Navarre, are very limited and they now retain responsibilities only for cultural matters and for roads. In practice, they provide services to Ayuntamientos with fewer than 20,000 inhabitants, managing local taxes, water supplies, and cultural and communications infrastructures.

At the local level, there are around 8,000 local authorities (Ayuntamientos) throughout Spain. Those with populations above 20,000 have considerable powers in the field of urban planning, which are expressly recognised in the Constitution. They also provide important services, such as water and electricity supply, sewage and rubbish collection, and can adopt certain initiatives in the field of local development and social policy. Apart from the responsibilities they exercise by law, some of these larger Ayuntamientos also coordinate with regional governments in the application of certain regional policies, to make the services they jointly provide, particularly in the area of social services, more direct and efficient. All this implies an expansion in local power to a point where some authors (Navarro Yáñez, 1999) identify a sort of new localism, which is creating an opportunity structure that allows people to participate in politics and, thereby, bring about a revitalisation of democracy at the local level. Since, however, this growth in local policy making is the result not of a downward process of transfer of powers from supra-municipal bodies but of initiatives upward from the Ayuntamientos, there is a great variation in the activities of local authorities in Spain. It is, therefore, impossible to define a standard political model for this fourth tier. In some areas, Ayuntamientos have coordinated their initiatives and established inter-local groupings (Mancomunidades) in order to carry out local policies more efficiently by working together and taking advantage of economies of scale. In other areas, such coordination is much less developed.

Administrative Structures for Agriculture and Rural Development

As far as agriculture and rural development are concerned, regional governments have assumed nearly all the powers that used to belong to the central government. The administrative structure, through which these powers are exercised, varies between regions, with each regional government adopting a system that best suits its needs, without reference to a common model.

Nevertheless, all regions share certain characteristics. Responsibilities for agriculture and rural development are generally concentrated in one regional government department. Although responsibilities for the environment and for town and country planning normally belong to departments other than those for agriculture, in some regions they are all concentrated in a single department. In all cases, the regional departments for agriculture and rural development, whether or not they have additional responsibilities, operate both centrally, from their headquarters in the capital city of each region, and from offices in the provincial capitals. In some regions, such as Andalusia, which has eight provinces, around 800 municipios and a population of 7.5 million, regional departments for agriculture and rural development have decentralised their administration to district (*comarca*) level, between the provincial and municipal levels, with an agricultural office in each district to assist farmers in their dealings with bureaucracy and administration. Dialogue with representative bodies, such as farmers' unions and federations of cooperatives, takes place at both the regional level and at the provincial and district level. In some regions, this dialogue takes place either within the framework of the Chambers of Agriculture – which come under the regional departments for agriculture and rural development and act as consultative councils in which farmers participate through democratically elected representatives[1] – or in the agricultural district offices.

Once responsibilities for agriculture and rural development have been transferred to the regional governments, the only function retained by central government is the coordination of negotiations with the EU, since member states are represented through their national governments regardless of their internal political and administrative organisation. In this respect, if we focus on the areas of agricultural structural policy and rural development, when the European Commission or the Council of Ministers of Agriculture approves a regulation, the Spanish government merely approves the corresponding national decree for its adoption. This decree is usually a more or less literal transcription of the EU regulation, which regional governments are left to apply according to their own rules, exploiting the room for manoeuvre that this allows them. Equally, when it comes to the coordination of proposals from regional governments, the national government, through the Minister for Agriculture, enters into discussion with them[2] in order to establish agreed criteria for a national proposal, which has then to be put to the EU. This is what happens, for example, when the Development Plan, which the Spanish government presents every four to six years to Brussels, is being prepared. Once it is approved by the European Commission and included in the European Community Support Frameworks, it provides the necessary basis for the approval of structural funds. The final project is the result of the coordination of 17 regional development plans, proposed by the different regional governments.

The provincial governments and Ayuntamientos do not participate directly in the preparation of policies for agriculture and rural development, but they are usually involved in the application of some development programmes. Some have, for example, participated directly in LEADER and PRODER programmes, as part of Local Action Groups, by making a financial contribution to the projects or by establishing complementary technical support networks, as we shall see later.

The Rural Restructuring Process

Recognising Rural Restructuring

As noted in the introduction, until the 1980s rural Spain was influenced strongly by the industrialisation process carried out in urban areas (Pérez Yruela, 1995). In short, persons of a productive age abandoned rural areas in search of better job opportunities. Those who stayed behind increased farming's productivity or took employment in non-farming jobs.

During the last two decades, some older people have returned to their rural places of origin after retirement from urban jobs. Young people are also moving to work or live in rural areas (García Sanz, 1996). These new tendencies, together with the fact that emigration from rural areas has been reduced considerably, seems to show that there is a revaluation of rural areas. The reasons for this cannot be examined here but are linked to housing being cheaper than in urban areas and the countryside offering a healthier and quieter environment. Improvement in roads and communications has also reduced the isolation of Spanish rural areas and attracted investment from entrepreneurs who want to take advantage of support policies implemented by local governments. Finally, jobs linked to the welfare state have allowed new professionals (mainly in the health or education services) to stay in rural areas.

The above changes have generated a strong theoretical debate within Spanish rural studies. This debate is well summarised in a collection that includes several articles on the main issues facing contemporary Spanish rural sociology (Gómez Benito and González Rodríguez, 1997). We may conclude that rural sociology's interests form two main groups: the farming production system and the characterisation of rural society. The first area gathers together analyses of agricultural modernisation, the agri-food system, the organisation of labour within the family farm, and the effects of public policies on agriculture. Debate is focused on the gap existing between Spanish agriculture and that of more modernised European countries, and on the need to reduce it through a new policy of agricultural modernisation. Some promote continued productivist modernisation, while others consider it more suitable to emphasise the ecological value of the Spanish countryside according to the sustainability paradigm.[3] The second area of interest includes issues such as the features and types of rural societies, and demographic and labour market changes. There are two basic controversies. The first concerns the issue of the existence or not of a rural society today. The demographic, economic and cultural changes in recent times make some authors, such as Camarero (1993) and Oliva (1995), conclude that the classic distinction between urban and rural society has been broken, and make them doubt of the usefulness of the concept 'rural society' as a category for analysis. In contrast, García Sanz (1996) is the best representative of those who favour continuing to use the concept of rural society, with both farmers and non-farmers as relevant actors. The second controversy relates to the best indicator to define a rural society. In the case of rural society existing as something specific, there is no agreement on the most adequate criteria to define it, but criteria include population density, number of inhabitants, certain cultural elements, some aspects linked to the social relations of its population (see Sancho Hazak, 1997). Although we cannot enter these debates here, the existence of a rural society will be acknowledged, if only in analytic terms, and it will encompass municipalities with less than 10,000 inhabitants.

Evidence of Rural Restructuring

To quantify the restructuring process being followed by Spanish rural society, the 17 regions in which Spain is divided administratively will be taken into account. The intensity and type of rural restructuring suggests a typology for these regions. The data used to construct this typology are somewhat old, because they are taken from quantitative analysis undertaken by García Sanz (1996) using the latest population Census available (from 1991).[4]

As shown in Table 8.1, the Spanish population which resides in rural areas numbers around ten million, or a quarter of the total population. Most of this rural population (70 per cent) is located in six of the regions: Andalusia, Castile-La Mancha, Castile-Leon, Catalonia, Valencia and Galicia. There are two further regions with high overall populations, but these have small rural populations: Madrid and Basque Country.

Table 8.1   Population per size of municipality, by regions in 1991

	Up to 10,000 inhabitants		Over 10,000 inhabitants		Total
	No.	%	No.	%	
Andalusia	1,667,258	24	5,251,369	76	6,918,627
Aragon	414,004	35	769,712	65	1,183,716
Asturias	160,500	15	930,292	85	1,090,792
Balearic Islands	169,145	24	535,406	76	704,551
Canary Islands	283,072	19	1,204,567	81	1,487,639
Cantabria	193,555	37	331,641	63	525,196
Castile-La Mancha	907,543	55	744,658	45	1,652,201
Castile-Leon	1,199,312	47	1,339,787	53	2,539,099
Catalonia	1,194,697	20	4,841,383	80	6,036,080
Valencia	873,143	23	2,972,633	77	3,845,776
Extremadura	614,287	58	442,470	42	1,056,757
Galicia	958,655	35	1,759,796	65	2,718,451
Madrid	233,236	5	4,693,932	95	4,927,168
Murcia	108,437	10	932,877	90	1,041,314
Navarre	246,858	48	270,486	52	517,344
Basque Country	396,130	19	1,702,215	81	2,098,345
La Rioja	109,976	42	152,650	58	262,626
Total	9,729,808	25	28,875,874	75	38,605,682

Source: elaboration of data from García Sanz, 1997, pp. 477–91.

The proportion of the rural population varies considerably across the regions, as is also clear from Table 8.1. Three main groups can be differentiated. The first are the most rural regions, those where very high percentages of the population (42–58 per cent) live in municipalities with less than 10,000 inhabitants: Castile-La Mancha, Castile-Leon, Extremadura, Navarre and La Rioja. A second group are the least rural regions, where only a small minority of the population (5–20 per cent) live in small municipalities: Asturias, Canary Islands, Catalonia, Madrid, Murcia and Basque Country. The remaining regions have intermediate levels of rural population (23–37 per cent). These differences in the proportion of the population rural in each region are due to many factors, such as the settlement pattern and the ways the population has adapted to it. Also significant are factors related to the rural restructuring process over the past 50 years, such as migration from areas where farming production was not very profitable and where there were no other jobs available.

Spanish population followed a clear tendency towards concentration in urban areas during the period between 1950 and 1991, illustrated in Table 8.2. Figures available distinguish three sizes of municipalities and, in this case, we consider as rural those municipalities with less of 2,000 inhabitants. Overall, bigger municipalities increased their population in these last decades, middle sized municipalities kept a stable population, and smaller municipalities witnessed a considerable decrease of inhabitants. This depopulation of smaller municipalities occurred in all regions, except in the Balearic Islands, but is most obvious in certain regions, notably Andalusia, Aragon, Asturias, Castile-La Mancha, Castile-Leon, Murcia, Basque Country and La Rioja. In these regions, their smaller municipalities saw a reduction in population by almost a half. It is in these regions where the rural restructuring process has been most intense.

The impact of restructuring in the rural areas does not depend exclusively on the evolution of its population in the last decades but also on the number of inhabitants left after depopulation. The future potential of regions which keep a high number of rural people despite having suffered net out-migration might be more optimistic than those where the population has been heavily reduced. In Castile-La Mancha, Castile-Leon and La Rioja, rural areas still account for a high proportion of the population despite large population losses. Elsewhere, heavy depopulation has considerably marginalised rural areas: Asturias, Murcia and Basque Country.

The decrease of the population in rural areas allows us to quantify the impact of rural restructuring, but it does not offer information on its characteristics. An analysis of the demographic and employment features of the rural population help us to understand the process further. Age structure is an indicator of the past, present and future potential of a region. This is shown in Table 8.3. In the rural municipalities of all Spanish regions the proportion of people aged above 65 years is higher than in those with more than 10,000

Table 8.2   Evolution of population between 1950 and 1991 per size of municipality, by regions (base 1950=100)

	Under 2,000 inhabitants	2,000–10,000 inhabitants	Over 10,000 inhabitants
Andalusia	53	84	198
Aragon	51	69	244
Asturias	56	218	331
Balearic Islands	110	153	207
Canary Islands	89	269	320
Cantabria	73	155	250
Castile-La Mancha	57	66	143
Castile-Leon	50	82	239
Catalonia	72	155	249
Valencia	65	88	289
Extremadura	93	52	123
Galicia	76	177	237
Madrid	86	137	283
Murcia	56	155	257
Navarre	64	108	391
Basque Country	55	163	328
La Rioja	56	86	240
Total	63	100	245

Source: García Sanz, 1997, p. 75.

inhabitants. Rural areas, therefore, offer a more limited socioeconomic potential than non-rural areas. This scarce potential seems especially evident in five regions: Aragon, Asturias, Castile-Leon, Galicia and La Rioja. In contrast, four other regions (Andalusia, the Canary Islands, Madrid and Murcia) display a rejuvenated structure to their rural population. In the Madrid region, the rural proportion of people aged under 14 years is even greater than in non-rural municipalities, confirming the tendency among some younger families to choose rural areas as places for living.

More information about the features of rural restructuring comes through analysis of the labour activity of the population in rural areas, shown in Table 8.4. The existence of non-farming employment, in particular, alters substantially traditional rural society. In contrast with the increasing predominance of the service sector in the economy overall, rural areas have a more balanced productive structure. Although the tertiary sector provides the most jobs (36 per cent of the employed population in 1991), other sectors also

Table 8.3   Percentage of inhabitants per age group and size of municipality, by regions in 1991

	0–14 years		15–64 years		65+ years	
	<10,001 persons	>10,000 persons	<10,001 persons	>10,000 persons	<10,001 persons	>10,000 persons
Andalusia	21	24	64	65	15	11
Aragon	14	17	63	67	23	16
Asturias	15	16	62	68	23	16
Balearic Islands	18	20	65	67	17	13
Canary Islands	21	22	67	69	12	9
Cantabria	18	19	65	67	17	14
Castile-La Mancha	18	22	62	66	20	12
Castile-Leon	15	19	63	67	22	14
Catalonia	18	18	66	68	16	14
Valencia	19	21	65	67	16	12
Extremadura	19	23	63	65	18	12
Galicia	15	19	63	68	22	13
Madrid	21	19	66	69	13	12
Murcia	22	23	65	65	13	12
Navarre	16	18	66	69	18	13
Basque Country	17	17	69	71	14	12
La Rioja	15	18	65	66	20	16
Total	18	20	64	68	18	12

Source: García Sanz, 1997, pp. 498–500.

account for significant percentages. Nonetheless, as agricultural activity only represents 27 per cent of the employed population, rural municipalities can no longer be catalogued as farming areas.

Once more, the structure of employment is not homogeneous across Spanish rural regions. Three main groups of rural society can be differentiated. First are those regions which maintain a clear agricultural component, such as Andalusia, Asturias, Extremadura and Galicia. In the rural areas of these regions, farming activity occupies between 34 per cent and 45 per cent of the employed population. Second are the rural municipalities where the service sector has acquired a clear predominance, notably the Balearic Islands, the Canary Islands and Madrid. Finally, there are those municipalities where industrial activity has a similar importance to the service sector. This occupational structure seems well represented in the rural areas of Catalonia, Navarre and the Basque Country. Municipalities with the last two types of

Table 8.4   Percentage of employed population per economic sector in municipalities with fewer than 10,000 inhabitants, by regions in 1991

	Economic sector			
	Primary	Industrial	Construction	Tertiary
Andalusia	45	12	13	30
Aragon	28	27	12	33
Asturias	34	19	11	36
Balearic Islands	9	14	17	60
Canary Islands	16	8	19	57
Cantabria	25	25	14	36
Castile-La Mancha	25	23	19	33
Castile-Leon	31	19	14	36
Catalonia	13	37	12	38
Valencia	19	31	14	36
Extremadura	41	12	15	32
Galicia	41	14	17	28
Madrid	7	25	18	50
Murcia	27	25	14	34
Navarre	15	37	12	36
Basque Country	8	42	9	41
La Rioja	24	32	12	32
Total	27	23	14	36

Source: García Sanz, 1997, p. 548.

employment structure are those where rural restructuring has most disengaged the rural population from its traditional farming base.

A Heterogenous Experience

The rural restructuring process in Spain is intense but not homogeneous. The reasons for this heterogeneity are multiple:

* rural areas are not uniform;
* globalisation's penetration is not identical in all areas;
* the ways the rural population react are not mimetic.

Thus, for example, while the rural areas of certain regions are made up by a multitude of small and dispersed villages, in others the population is concentrated in a single village. Similarly, whilst in some municipalities there are non-farming jobs nearby, in others the distance to the work place is so great that it becomes necessary to abandon a rural place of residence.

To appreciate better the intensity and the features of restructuring in Spain, two indicators from the previous analysis are combined. The first one, the loss of rural population in recent years, reflects the intensity of restructuring. The second indicator, the economic structure of rural areas, measures the characteristics of the restructuring. The typology resulting from crossing both indicators, shown in Table 8.5, may be imprecise, but it provides a first attempt to understand the heterogeneity of rural restructuring.

Table 8.5   A classification of rural Spain by population change 1950–91 and economic sector profile, by region in 1991

	Population loss	Main economic sector	Rural population	
			No.	%
Andalusia				
Asturias	High	Agriculture	1,827,758	19
Aragon				
Castile-La Mancha	High	Equilibrium	2,739,272	28
Castile-Leon				
Murcia				
La Rioja				
Basque Country	High	Industry and services	396,130	4
	High	Services	0	0
Galicia	Medium	Agriculture	958,655	10
Cantabria				
Valencia	Medium	Equilibrium	1,066,698	11
Catalonia				
Navarre	Medium	Industry and services	1,441,555	15
	Medium	Services	0	0
Extremadura	Low	Agriculture	614,287	6
	Low	Equilibrium	0	0
	Low	Industry and services	0	0
Balearic Islands				
Canary Islands				
Madrid	Low	Services	685,453	7
Total	–	–	9,729,808	100

According to both indicators, the most common impact of rural restructuring in Spain is the loss of a large part of the population and the moderate growth of the non-farming sector. This tendency is most common in the rural areas of Aragon, Castile-La Mancha, Castile-Leon, Murcia and La Rioja, accounting for 28 per cent of the Spanish rural population. A second pattern of restructuring is followed by two rural areas from very different regions: Andalusia and Asturias. Both suffered high population loss but they have retained the economic importance of the agricultural sector. Around 19 per cent of the Spanish rural population are found in these regions. A third important way in which rural restructuring has manifested itself has been through moderate loss of population but with a great development of jobs related to the industrial and service sectors. This is observed in the rural areas of Catalonia and Navarre, which have 15 per cent of the Spanish rural population. In general, some kind of negative relationship can be found between the evolution of rural population and the importance of the farming sector. In rural municipalities with high population loss the service sector has not developed, and the importance of the industrial and service sectors is low (with the exception of the Basque Country). In contrast, regions with low population loss tend to have developed their service sector (with the exception of Extremadura).

Rurality has changed in Spain over recent years (Pérez Yruela, 1995). Industrialisation of the economy in the 1950s favoured the migration of rural population to urban areas in order to increase incomes. Rural society was perceived as a space linked to traditional farming practices that were unable to provide a decent living. The decreasing proportionate weight of agriculture in the economy and of the rural population in the total population seemed inevitable and desirable. Yet, at the beginning of the 1980s, some negative consequences of this process started to emerge. Cities had become massive and problems derived from poverty and urban living were more evident. On the other hand, some advantages of rural areas were realised, such as their cultural and ecological heritage. There were new possibilities for leisure and even for having a better lifestyle in rural Spain. Rural society is seen nowadays as a space for consumption more than as a place for production. Moreover, since rural society depends more than ever on the needs of urban society, its villages' relative locations are very important for their future roles. Those close and well linked to urban centres may turn into 'sleeping villages', dominated by urban commuters and second homes. Villages poorly linked to these centres but with cultural or ecological resources may become new tourist destinations. Rural areas with poor roads, few resources to offer, or a weak agrarian or industrial economy may be condemned to disappear.

Rural Development Initiatives

In 1994 a new Community Support Framework (1994–99) was approved, adding a second rural development programme (for Objective 1 and 5b areas in Spain) to the European Commission LEADER II initiative. Included within Agriculture and Rural Development was a programme for rural development – Programa Operativo de Desarrollo y Diversificación Económica de las Zonas Rurales (PRODER). From a conceptual point of view, the PRODER programme should have followed a top-down logic, with the strong government intervention apparent in other operative programmes. In practice, it was adopted according to a bottom-up logic, typical of the LEADER initiative. Indeed, the PRODER programme was implemented to benefit rural areas that had been unable to join LEADER II, with criteria very similar to those for the latter for the selection of areas and the management of programmes. LEADER II and PRODER complement each other in extending throughout the country a network of 233 Rural Development Groups[5] (133 in the LEADER programme, 100 in PRODER) in areas with fewer than 100,000 inhabitants.

The establishment of programmes under LEADER II and PRODER followed the same four-phase pattern. The first step was taken by the regional agricultural departments, whose task was to select Rural Development Groups (RDGs). A distinction was made between groups who had to follow a six month skills-training course, since they were considered neither sufficiently experienced nor trained to embark on a programme of local development, and those that could immediately start carrying out the programme. Generally speaking, Ayuntamientos played an important role in encouraging the establishment of RDGs and presenting the programme or starting skills-training. They brought together other actors involved in local development, such as business groups, cooperatives, unions, cultural organisations and financial bodies. Technical criteria, such as the interest, coherence and viability of innovative rural programmes, were not the only ones used in selecting RDGs. Regional agricultural departments also took into account the risk of concentrating programmes in one area to the detriment of others and sought balance within each region, so that all areas benefited from some programme. Equity was also a factor, with priority given to less favoured areas, even though the quality of the programmes presented might have been low. There were cases, too, where certain mayors/council leaders sought to take advantage of their political affinity with the regional government.

In the second phase, an organisational structure was established to implement approved rural development programmes, either directly or after the skills-training course. Responsibility here lay with the RDG. In order to carry out their programme, they created a Centre for Rural Development (CEDER) in each district, with a Director and technical staff financed by LEADER or PRODER funds. The CEDER was the technical structure of each RDG, but it did not have decision-making functions. In order to carry out these management functions, each RDG was organised according to different legal formulae as civic groups, commercial organisations, consortiums, foundations, etc.

The third phase saw the approval of different projects presented by those with an interest in rural development in each area. CEDER technical staff evaluated each project before it was accepted by the RDG's management body. There was quite an intense relationship, in this regard, between the CEDER (whose role in providing technical advice and developing sociocultural awareness was significant), each RDG's management body (which defined the criteria for approving projects and established priorities for rural development) and the agricultural departments of each regional government (responsible, through their administrative services, for giving final approval to projects sent from RDGs and for providing the necessary finance).

In the final phase, the LEADER and PRODER programmes were evaluated by external advisers, acting for each regional agricultural department. They assessed key aspects, such as the level to which the Programme of Rural Innovation was implemented in each area, the number of projects presented and approved, the volume of investment, the amount of direct and indirect employment generated, and public opinion on the effects of the programmes.

The LEADER II and PRODER Programmes in Andalusia

The case of Andalusia is interesting to analyse for several reasons. Firstly, it is the Spanish region where the most RDGs were created and where these groups coordinated their strategies through an umbrella organisational structure (Asociación Rural de Andalucía, ARA). Secondly, Andalusia is participating in the debate on rural development with important contributions by economists and sociologists, mainly from the university and the research centres of Cordoba. Thirdly, the regional government elaborated an important legislative framework to regulate the RDGs. Finally, Andalusia was the first Spanish region where LEADER II and PRODER were evaluated externally.

In Andalusia, 666 of the 770 municipios (85.4 per cent of the total number, 88.6 per cent of the land area, 41.8 per cent of the population) were affected by rural development activities under the two programmes during the period

1994–99. If we take out the eight provincial capitals, the coastal areas and the large municipios (which are basically urban and fall within the sphere of influence of the capitals), practically the whole of rural Andalusia was involved in rural development programmes. Within the framework of these programmes, 49 Rural Development Groups were established, 22 in LEADER II and 27 in PRODER. This is illustrated in Figure 8.2.

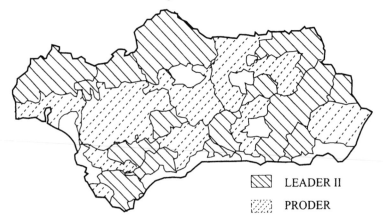

LEADER II

PRODER

Figure 8.2   Distribution by area of the LEADER II and PRODER programmes in Andalusia

Source: *Guía del Desarrollo Rural*, Empresa Pública DAP, Junta de Andalucía, Sevilla 1999.

Most RDGs were established in areas with an historically strong, shared cultural identity, or where the sense of identity was not so strong but where there were shared territorial features. This could be explained by the bottom-up logic of LEADER and PRODER, which encouraged relations between local actors sharing common problems. The creation of an RDG had a stimulating effect on the social capital of the local community. [6] As a result of the LEADER programme, 131 new organisations sprung up, 66 of them business groups, 60 women's, cultural, youth and environmental groups, and five political. The number of members in existing groups also increased. Another effect was the creation both of consultative committees in each RDG and of a network of technicians and rural development workers in each CEDER.

As for financial investment, 32,225 million pesetas were invested in LEADER II and 25,966 million pesetas in PRODER. The origin of the funds invested is shown in Table 8.6. Private initiative was clearly important, representing more than 40 per cent of total investment in LEADER and nearly 30 per cent in PRODER, with the rest coming from EU funds (according to the rules for co-financing Objective 1 and 5b areas) and from national, regional and local governments. Projects receiving the largest subsidies were rural tourism (accounting for more than 50 per cent in both programmes), the agri-food industry (with 30 per cent and 35 per cent in LEADER and PRODER, respectively, going to small companies), dealing in local products, and projects concerned with training and support for small and medium size businesses.

Table 8.6   The origin of funds invested in LEADER II and PRODER in Andalusia (million pesetas)

	LEADER II	PRODER
European Union	12,900	11,600
Central government	2,800	896
Regional government	2,800	2,100
Local authority	825	2,100
Private initiative	12,900	9,270
Total	32,225	25,966

After nearly a decade of experience in rural development, the Andalusia regional government regards the existing 49 RDGs as an important resource for the region, which it should maintain and encourage. It therefore included a major proposal for rural development in the next Development Plan presented by the Spanish government to the European Commission. This will make up the Community Support Framework (CSF) for the period 2000–06. This proposal foresees the provision of resources to finance development programmes submitted by the RDGs. For the period 2000–06, they will have at their disposal 72,000 million pesetas from public funds, provided by the EU (through the CSF) and the regional and national governments, to finance privately initiated projects.

In conclusion, Andalusia is an interesting case study on the implementation of rural development policy. Several factors explain the success of the policy. First, there were political factors, such as the presence of left-wing parties in both regional and local governments for 20 years. Second, cultural factors such as the emergence of a new local-oriented values system, favoured the development of rural communities. Third, there were socioeconomic factors, such the existence of a dynamic network of middle size municipalities (10,000–15,000 inhabitants) characterised by the combination of urban and agricultural activities. Some authors (for example, López Casero, 1997) have called these *agrociudades*.

The Effects of the LEADER and PRODER Programmes on Spain's Rural Areas

Analysts generally agree that the adoption of the LEADER and PRODER programmes has led to the establishment of a development model that, with one of its main elements being the participation of social actors, has invigorated and given voice to many rural areas in Spain (Esparcia et al., 1999; Esparcia, 2000). Before these rural development programmes, the dynamic of many Spanish rural areas was blocked, due to their dependence on agricultural activity. For many people, especially women and the young, migration to towns was the only option for upward mobility. Since the mid-1980s, however, both local policies carried out by the new democratic municipalities and EU development programmes have presented new opportunities to the non-agricultural population (Moyano, 1999; Navarro Yáñez, 1997). Although there have been variations in the ability of different areas to take advantage of these opportunities, Spanish rural society has demonstrated that it is capable of responding and that it can organise itself and adopt initiatives that differ from those traditionally associated with agriculture.

The creation of RDGs has been a fundamental element in this change, providing rural areas with a new participatory opportunity structure from which they can direct their own development. This process has given rise to a rural development model whose most important characteristics we now examine. This discussion is based on the intermediate evaluation of the LEADER II programmes in Andalusia by the Enterprise for Agricultural and Fish Development (DAP), in collaboration with the Research Institute for Advancing Social Studies (IESA, part of the Spanish Council for Scientific Research) (DAP, 1999). In addition, there is more general information on the LEADER and PRODER programmes from technicians of the Ministry of Agriculture, Fisheries and Food (Alvarez, 1998, 1999).

The Emergence of a New Local Identity

In accordance with the established territorial criteria and with Objectives 1 and 5b, most of the LEADER and PRODER programmes were in places where social organisation was poor, which might be expected to be a serious handicap to generating initiatives. In these areas, the farming sector was not dynamic enough to encourage economic activity, and the agricultural population was aged. In practice, however, in areas with a strong cultural identity, or where they were able to recover this identity as another element in their development strategy, and where there was experience in working together, conditions for change were more likely to be created. During the first phase of the rural development programmes, the public sector generally assumed the role of instigator and took on the responsibility of promotion and the provision of technical support. In most RDGs, the role of Ayuntamientos, through their mayors, was decisive, notwithstanding the need for frequent intervention by the regional agricultural departments to resolve different interests and establish joint criteria where there was a lack of consensus at the local level. From the point of view of territorial identity, the programmes contributed to the definition of an inter-local community (*comarca*) where there once was only a collection of small villages, absorbed in local conflicts and ancient disputes among neighbours. The programmes made it possible for village leaders, who would never have sat down and talked to each other before, to start to think together about the future of the area and to look for common solutions. After ten years of LEADER in some areas, the population felt a wider sense of belonging. Thus, in a 1997 survey (Moyano and Pérez Yruela, 1999), more than 50 per cent of Andalusian people felt a belonging to both a local (pueblo) and a supra-local (*comarca*) community.

Social Participation

Establishing cooperation proved to be a complex process, requiring frequent local and area meetings and the sociocultural stimulus provided daily by CEDER. One of the great challenges from the outset was the participation of social actors. The most effective types of participation proved to be those that stimulated horizontal cooperation through the establishment of consultative committees, which made it easier for different social actors in the area to meet together properly. Generally speaking, LEADER and PRODER programmes were perceived by local people as having contributed to an improvement in social cohesion in rural areas, by creating a balance between the public and private sectors, and by improving the representative nature of social, economic and political organisations. In this sense, LEADER and PRODER encouraged the building of social capital in many rural areas.[7]

Of the different collectives involved in the rural development process, that of business was the most influential, explained by the nature of the development programmes. However, special mention should be made of women's collectives, where great advances were made in setting up groups, with subsequent participation in development activities and rural social life. As for young people, although these programmes in many areas made them aware of the possibilities offered by their own local resources, their involvement was not very great. This is one of the tasks pending that people feel should be given greater attention. Encouragement should be given to the relationship between the CEDERs and the education system at local level, particularly in technical education, where there is a potential that, if well channelled, could be a source of enterprising initiatives for development.

The total lack of farmer participation deserves particular mention. In the light of the projects presented and the composition of the RDGs, these areas can be said to have 'rural development without farmers'. Farmers did not feel any sense of identification with the programmes, nor, with a few exceptions, did they show any interest in getting involved in the initiatives and projects. Their principal concern, and that of their unions, continued to be Common Agricultural Policy direct payments and subsidies. Although some unions and cooperatives took part formally in the initial phase of establishing RDGs, their involvement in getting development programmes going was marginal. There is the risk of a gap opening between farmers and non-farmers in rural areas. Farmers seemed to belong to a corporatist policy community only interested in direct payments and not in the general problems of rural society.

Management and Technical Structure for Development

As far as the legal formula adopted by the RDG's management body was concerned, the most common was that of a commercial organisation (14 of the LEADER II groups in Andalusia), followed by a civic association (seven) and a consortium (only one). The civic association seems to have generated a greater sense of popular identification with the development programme, and a feeling that it was something that 'belonged to them'. This is in spite of the fact that it created management problems, owing to the difficulty of reaching group decisions. In reality, in RDGs that chose this form of civil association, the rhetoric of participation did not always result in the actual involvement of the different collectives. There was less identification with the commercial organisation formula. Here, the programme was popularly seen as 'something foreign', coming from outside, and there were accusations of entrepreneurial elitism. Nonetheless, this formula showed itself to be more efficient in the management of resources and in the conduct of projects. Finally, the consortium formula displayed an interesting potential to allow collaboration between public and non-public institutions at the local level in the implementation of development programmes.

Regarding the technical structure, one of the great achievements of the LEADER and PRODER programmes was the establishment of a multi-disciplinary structure (CEDER), with training and experience in rural development. This was particularly valued by local people for the quality of its personnel and commitment to the area. It drew out a hitherto unknown development culture in many areas. The CEDER provided a 'single window' to simplify bureaucratic procedures, by dealing with requests from potential beneficiaries and rerouting projects the programme could not subsidise to others that provided support. RDGs, through the CEDERs, were able to attract other public funds and bring together all those interested in a common development project. In some areas, the CEDER's technical structure was complemented by technicians from other development programmes already active in the same area – local development workers belonging to provincial government programmes, for example – and by networks of volunteers. The LEADER and PRODER programmes, therefore, encouraged a synergy in favour of local/rural development. Where this happened, optimum use was made of resources and administrative procedures were improved, avoiding duplication and allowing proper attention to be given to beneficiaries and to the dissemination of the programmes. However, where such synergy did not occur, competition among technicians from different development programmes had negative effects.

The Nature of Activities Implemented

The majority of RDGs engaged in activities that made a significant contribution to the creation of businesses, the generation of employment, the reassessment of local resources, and the provision of services to people. Many projects were directed towards the creation of hotel and restaurant infrastructure, the modernisation of agri-food industries, and the provision of local services. Some RDGs promoted collective or sector projects, entailing a closer relationship between sponsors. There was a greater interaction between different sectors in projects related to promotion. There were many examples of markets, agricultural fairs, contests, etc., where activity was directed to the recovery of traditions together with the promotion of local products.

In short, although RDGs began to operate without a clear strategy for intervention or development, this changed with time, as collective projects and innovative activities with a demonstrable effect became increasingly frequent. However, since no clear definition of the concept of innovation exists, its implementation in practice varied according to the socioeconomic and geographical characteristics of different areas. While some RDGs regarded as innovative any project that implied changing a negative trend (for example, depopulation, unemployment, the black economy), for others it meant covering a shortfall in products and services. The more advanced RDGs used environmental and cultural criteria, such as creating awareness, environmentally sensitive production or the recovery of traditions, to distinguish innovative projects. The formation of training activities was, however, clearly an innovation. The establishment of structures for the design, follow-up and evaluation of training activities in which businesses and government officials took part, or the realisation of specific activities for groups with particular difficulties of access to the labour market (especially young people and women), involved radical innovation with regard to what had gone before in rural Spain. The most innovative aspect, nonetheless, was undoubtedly the philosophy of the LEADER and PRODER programmes, which imposed a systematic approach to work – following up projects, making sure they are suitable – promoted participation, and encouraged a change of outlook on the options available. In short, innovation involved qualitative change, which encouraged a new social articulation and the diversification of the rural economy.

The Demonstrative Effect

The demonstrative effect, understood as giving projects a certain 'visibility' so that they can serve as an example to other initiatives, was not generally achieved. Above all, this was due to the lack of a proper strategic plan.[8] Demonstration was most effective in the rural tourist industry, where the maintenance of an architectural tradition, gastronomic specialisation, or improvements in the quality of service to the client spread throughout the area very easily. Even the restoration of country houses was useful as a reference point for other accommodation in the area. The promotion of women-led businesses was another good example, especially in sectors such as textiles, where an unregulated economy is an important feature. Other projects with an important demonstrative effect were those directed at the diversification of the primary sector, especially organic production, the labelling of agri-food products and the modernisation of forestry companies. Certainly, even though few projects might be undertaken, the popular view was that those that were approved should be done well, so as to serve as a stimulus for future entrepreneurs. Overall, the LEADER and PRODER programmes, although less beneficial in terms of economic resources than initially hoped, were felt to have had an important demonstrative effect locally. They provided an opportunity for attracting the interest of regional and European institutions to the area, not least by exploiting the value of 'the endogenous' as a basis for interchange. This may open up new development possibilities.

Gender Issues

Progress by women in assuming a more active role in the dynamics of society and the economy is fundamental to the future of rural Spain. Yet, in spite of such recognition, there was no general incorporation of gender issues at the horizontal level in the activities of the RDGs, even though there were some concrete actions which contributed to the improvement of the initial situation. Data availability limit analysis of the composition of the RDGs but it is clear that there were very few women involved in decision-making. Where they did exist they were usually representatives of women's organisations or businesswomen. The situation in the CEDERs was similar. It is clear that both the RDGs and the CEDERs reflected the current situation in rural society. Nonetheless, although much more needs to be done in this respect, the incorporation of women, both in the technical teams and in decision-making, ought to make improvements easier. Having the opportunity to take part in events, to travel, to train and, above all, to make decisions, encouraged many women to improve their own situation. Women were increasingly setting up businesses and leaving precarious employment thanks to the opportunities offered by rural development programmes (Navarro Yáñez, 1999).

Extra-local Networking

RDGs were generally aware of the importance of inter-regional cooperation. Many established partnerships and maintained ongoing relationships. There were different networks of RDGs with a particular interest, such as those in mining valleys or those with Natural Parks in their area. Cooperation was valued especially highly for any increase it might bring to the size of the RDG's territory. Many RDGs valued positively the European Association for Information and Local Development, with connection to this association and the recent creation of the Spanish Unit of the European Observatory seen as potentially useful tools for networking. The most frequent issues involved in the networking process concerned rural tourism, the recovery of cultural heritage, crafts and the development of factors that facilitated the promotion and joint marketing of quality agri-food products, such as brands and Protected Designation of Origin. Those involved in networking, almost without exception, were CEDER technical staff and representatives of local government. The business sector took part only in exceptional cases.

The Role of Public and Non-Public Institutions

Local administrations (Ayuntamientos, provincial governments and district councils) played different roles in the process of implementing the LEADER and PRODER programmes. During the initial skills-training period, when there was no previous experience of participatory procedures and much less of development involving numerous social and economic agents, local authorities played a leading role in setting up the RDGs and in implementing the development strategies. In some areas, the level of socioeconomic organisation was very low and only local authorities were able to provide a unifying force for these initiatives. In the second phase, once the programmes were operating, local authorities played a less active role. It was generally felt that Ayuntamientos should have effective involvement in LEADER and PRODER, not as managers nor as beneficiaries but as supporters and providers of logistic infrastructure for the CEDERs. Provincial governments were criticised for their lack of coordination between local development workers and the technical staff of the CEDERs. They were seen as contributing to the proliferation of local development bodies in rural areas without any more efficient service in return. With their authority as an association of local authorities, it was advocated that they coordinate and direct local development services in their provinces.

At the regional administrative level, most RDGs indicated that the central services of the regional agricultural departments were very late in replying to their requests. RDGs demanded greater clarification of the eligibility criteria used by the regional government. Some saw a need to adapt criteria to the characteristics of each area, and to give the RDGs greater freedom in decision-making. Another major problem that RDGs had with the authorities was the delay in providing funds, which caused serious financial difficulties, particularly for those with a high level of committed spending.

Financial institutions generally confined themselves to granting credit to projects or to the CEDERs themselves, but had not established a stable strategy for the development model in use. They were usually passive by nature, often merely taking in savings, and were insensitive to the needs of development programmes that did not rely most importantly on public funding. This concerned not only the provision of bridging finance for projects when subsidies were delayed, but also in depth involvement in development and in the placing of their specialised resources at the service of economic and social agents in each area. To the extent that they achieve this, their future in the area will be assured. The future of development in many areas is highly dependent upon the level of involvement of these bodies, particularly the savings banks.

Social Change and the Redefinition of Rurality in Spain

Until recently, the crisis in rural areas and the problems that threaten the farming population were often spoken of as though they were identical. In reality, however, the non-agricultural rural population has rarely been asked about how it perceives changes in agriculture and rural areas. Discussion about this ongoing change has been stamped, above all, by the agrarian background of many scientific researchers and technicians, the majority of whom come from university schools of agriculture or departments of the Ministry of Agriculture itself. Farmers have been the privileged reference group when explaining change in Spanish rural society and, not surprisingly, it has been their perception of crisis that has dominated research.

In the last few years, a new generation of research workers, sociologists, geographers and anthropologists has emerged from different university faculties. This generation does not have a background in agriculture and they are beginning to study change in rural society from the perspective of groups unrelated to farming.[9] The importance of their research is that it reveals a perception of change very different from that of farmers, unaccompanied by the crisis of identity that the latter normally experience. For many groups in the non-agricultural population, the present process of change offers considerable opportunities for invigorating rural life and improving the use of rural space and the land, in accordance with society's new demands.

Programmes for local/rural development, particularly those channelled through LEADER and PRODER, have encouraged the emergence of new leaders in the economic and social life of Spanish rural communities. New commercial initiatives supported by these programmes, as well as the proliferation of technical staff and development workers, are bringing a dynamism to rural areas that allows them to perceive the processes of change in a different way from farmers. Until recently they worked separately, confined within the limits of their own development programmes, but now they are beginning to organise themselves into associations, not only to share experiences but also to embark on wider activities. They are beginning to take part in national and international forums where the content of rural development policies is decided. They are critical of the agricultural bias of rural development policy and propose that it should no longer be channelled through departments of agriculture. Instead, it should be implemented by interdepartmental organisations of a horizontal nature, with the ability to integrate different activities. They do not accept, for example, that LEADER should continue to be channelled through the Ministry of Agriculture and the regional agricultural departments when the majority of its activities are non-agricultural.

Other actors emerging under the umbrella of the welfare state's health, education and social services are also playing a dynamic part in defining the future of rural areas in a way that is very different from the traditional view. They are frequently becoming directly involved in development programmes, whether individually or through the institutions to which they belong. Their presence in villages is increasing, particularly with improvements in communications and quality of life, and they are choosing to live where they work rather than leave rural Spain at the end of the working day, which was the norm until very recently. The importance of these sectors cannot be ignored, inasmuch as the impact on rural areas of education and health policies, which decide where schools and health centres should be located, is often greater than that of rural development programmes.

Lastly, the attention paid to leisure and recreation activities in rural areas also encourages the presence of the urban population, whether for holidays, weekends, excursions, walking or field sports (García Sanz, 1996). Besides reviving rural traditions, they also bring with them lifestyles that are typically urban: young people going out at night, discotheques, the large-scale use of cars. Young people adopting new types of agriculture, such as organic, are often members of local environmental groups. They bring a 'non-farming' attitude towards natural resources, distancing themselves from the traditional discourse of farmers and may cause internal rifts in the local farming sector.

For all of these groups, the present context of change offers opportunities for invigorating rural society and influencing local decision-making, not least by taking part in local politics. Increasingly, one finds among local councillors in rural Spain, doctors, teachers, social workers and environmentalists. These people are becoming part of local elites, together with other professionals and business people from non-agricultural sectors (Moyano, 1999). The disjointed and ill-organised response of these groups to the new framework of opportunities, however, demonstrates the diversity of interests in rural areas today. Social scientists must continue to research this diversity in order to broaden our understanding of the social and economic dynamics of rural society in Spain. In this context, research workers have before them an interesting medium through which to study whether a new rural identity is emerging. This medium is not exclusively agricultural. It is a synthesis of the different activities and professions with particular connection to an area which are coalescing in the act of development in small and medium sized nuclei of population. What must be clarified is whether this movement is strong enough to allow us to speak of the existence of a new rural identity or whether, on the contrary, we have numerous unconnected identities, with no sense of belonging to a single community or of having any shared interests.

Conclusions

According to Esparcia (2000, p. 206), LEADER and PRODER programmes have had

> a significant impact in Spain. At the very least, [they have] helped to raise awareness among politicians and professionals of the dynamics of rural areas. Private sector investment has been mobilised and some employment created, although little is known about the survival rates of projects.

The creation of RDGs has been fundamental in this impact, providing rural areas with a new participatory opportunity structure from which they can direct their own development. This process has given rise to a rural development model that has encouraged territorial identity and social participation giving voice to new actors, and has favoured economic diversification.

From the point of view of territorial identity, development programmes have contributed to the definition of an supra-local community (comarca) where there once was only a collection of small villages, absorbed in parochial disputes. The LEADER and PRODER programmes have also stimulated horizontal cooperation through the establishment of consultative committees. These have made it easier for the different social actors in an area to meet together constructively, by creating a balance between the public and private sectors and by improving the representative nature of social, economic and political organisations. In this sense, the development programmes have encouraged the building of social capital in rural Spain, not only from the point of view of associative density but also in terms of the trust generated amongst the population. They have contributed to the emergence of new social actors (women, young people, professionals) as protagonists of rural areas who have demonstrated more initiative and greater capacity of mobilisation than the traditional agricultural community.

Regarding the diversification of activities in rural areas, the impact of LEADER and PRODER programmes has been less strong than initially hoped. Perhaps, the lack of efficient strategic plans to guide the actions of RDGs helps to explain this fact. However, development programmes have had an important demonstrative effect locally. They have provided an opportunity for attracting the interest of regional institutions to the area, by exploiting the value of 'the endogenous' as a basis for interchange, which may open up new development possibilities. The most innovative aspect has been the philosophy of the LEADER and PRODER programmes, which have imposed a systematic approach to work, promoted participation and encouraged a change in perspective. A qualitative change has encouraged social re-articulation and economic diversification.

Finally, the implementation of local-oriented development programmes has revealed their limitations. In areas which have had LEADER and PRODER programmes, it is clear that the dynamics of development have reached such a level that local people have started to become aware of the limitations of these programmes in solving the challenges confronting them. There are problems of road infrastructure, sanitation, communications, etc. that require a qualitative leap in the programming of development. In these cases, local authorities and population become aware of the need of regional strategic plans beyond the local approach of both LEADER and PRODER.

Notes

1   Chambers of Agriculture belonged to the old Francoist corporatism and now come under the regional governments. During the last 10 years, according to criteria established by each regional government, they have reorganised and undergone a process of internal democratisation, with the introduction of elections for the appointment of members. During Franco's time, Chambers of Agriculture provided a great many services to farmers, but they have now lost most of these old responsibilities and act merely as consultative bodies. There are considerable regional differences in their level of reorganisation and democratisation. Some are close to extinction, waiting only for their officers to retire, while others have undergone a revival, introduced democratic elections and appointed new technical staff.

2   Every year, three or four regional meetings are called by the Minister of Agriculture to integrate the proposals of regional governments. Regional governments are represented by the most important authorities of their agricultural departments.

3   On this debate see Regidor (2000); on the position of farmers' unions see Moyano (1995) and Garrido Fernández (1999).
4   In this analysis, the provinces of Ceuta and Melilla are excluded because of the lack of information, but this exclusion is not, however, relevant for our study since both provinces are quite urban.
5   We refer in the text to Rural Development Groups as a generic title for groups that take the initiative and carry out programmes for rural development, whether within the framework of LEADER or PRODER. In LEADER terminology, these groups are known as Local Action Groups.
6   This notion of social capital is based on the definition by Putnam (1993) who emphasises the associative density at local community. However, other authors consider different dimensions more relevant to define the notion of social capital (Woolcock, 1998).
7   Here, we use the notion of social capital emphasising the dimension of trust.
8   Although one of the conditions demanded from RDGs, before being given a LEADER or PRODER programme, was the presentation of a strategic plan, such plans have not, in reality, provided a sufficiently solid and coherent development strategy.
9   For some decades, Spanish rural sociologists were integrated with agricultural economists in the Spanish Association of Agricultural Sociology and Economics. All perceived Spanish rural society as if it were only based on agriculture, with farmers as the dominant social group. Five years ago, when this association centred their activities only on agricultural economics, rural sociologists came together in a new research committee created within the Spanish Federation of Sociology (FES). This committee has since represented rural sociologists, coordinating the triennial workshop on Rural Sociology at the Spanish Congress for Sociology.

References

Alvarez, J. (1998), 'El Programa LEADER II en España', in *Actualidad LEADER Revista de Desarrollo Rural*, Unidad Española del Observatorio Europeo Leader Madrid, pp. 12–5.
Alvarez, J. (1999), *PRODER: Programa de Desarrollo y Diversificación Económica de Zonas Rurale*, paper presented to the Observatorio Europeo LEADER Seminario: integrar las eneseñanzas de LEADER en las políticas rurales, Valencia.
Camarero, L.A. (1993), *Del Éxodo Rural y del Exodo Urbano. Ocaso y Renacimiento de los Asentamientos Rurales en España*, Ministerio de Agricultura, Pesca y Alimentación, Madrid, Serie Estudios, No. 81.
DAP (1999), *Evaluación Intermedia de la Iniciativa LEADER II en Andalucía*, DAP, Sevilla.
Esparcia, J. (2000) 'The LEADER Programme and the rise of rural development in Spain', *Sociologia Ruralis*, vol. 40, pp. 200–8.
Esparcia, J., Noguera, J. and Buciega, A. (1999), 'Emerging Community Development in Spain. in E. Westholm, M. Moseley and N. Stenlas (eds), *Local Partnerships and Rural Development in Europe: a Literature Review of Practice and Theory*, Dalarna Research Institute, Falun, and Cheltenham and Gloucester College of Higher Education: Countryside and Community Research Unit.
García Sanz, B. (1996), *La Sociedad Rural ante el Siglo XXI*, Ministerio de Agricultura, Pesca y Alimentación, Madrid, Serie Estudios, No. 125.

Garrido Fernández, F. (1999), *Social Actors and the Implementation of the EU Agri-environmental Policy in Spain*, paper presented to the XVIII European Congress for Rural Sociology, Lund.

Giménez Guerrero, M. (1998), *Los Aspectos Socioeconómicos de los Modelos y Experiencias de Desarrollo Local en España y Otros Países Europeos*, Doctoral Thesis, University of Córdoba.

Gómez Benito, C. and González Rodríguez, J.J. (eds) (1997), *Agricultura y Sociedad en la España Contemporánea*, Centro de Investigaciones Sociológicas and Ministerio de Agricultura, Pesca y Alimentación, Madrid.

López Casero, F. (1997), 'Identidad, Estructura Social y Desarrollo Local. Redefinición del Pueblo, con Referencia Especial a las Agrociudades,' in C. Gómez Benito and J.J. González Rodríguez (eds), *Agricultura y Sociedad en la España Contemporánea*, Centro de Investigaciones Sociológicas and Ministerio de Agricultura, Pesca y Alimentación, Madrid, pp. 673–704.

Moyano, E. (1995), 'Farmers' Unions and the Restructuring of European Agriculture', *Sociologia Ruralis*, vol. 35, pp. 348–66.

Moyano, E. (1999), *Processes of Changes in the Spanish Rural Society. Plurality of Interests and New Opportunity Structures*, paper presented to the XVIII European Congress for Rural Sociology, Lund.

Moyano, E. and M. Pérez Yruela (1999), *Informe Social de Andalucía (1978–98). Dos Décadas de Cambio Social*, Federación de Cajas de Ahorros de Andalucía, Sevilla.

Navarro Yáñez, C. (1997), *Innovación Democrática en el Sur de Europa*, doctoral thesis, University of Complutense, Madrid.

Navarro Yáñez, C. (1999), 'Women and Social Mobility in Rural Spain', *Sociologia Ruralis*, vol. 39, pp. 222–35.

Oliva, J. (1995), *Mercado de Trabajo y Reestructuración Rural. Una Aproximación al Caso Castellano-Manchego*, Ministerio de Agricultura, Pesca y Alimentación, Madrid, Serie Estudios, No. 98.

Pérez Yruela, M. (1995), 'Spanish Rural Society in Transition', *Sociologia Ruralis*, vol. 35, pp. 276–97.

Putnam, R. (1993), *Making Democracy Work*, Princeton University Press, Princeton.

Regidor, J.G. (2000), *El Futuro del Medio Rural en España*, Consejo Económico y Social, Colección Estudios, Madrid.

Sancho Hazak, R. (1997), 'Las Políticas Socioestructurales en la Modernización del Mundo Rural', in C. Gómez Benito and J.J. González Rodríguez (eds), *Agricultura y Sociedad en la España Contemporánea*, Centro de Investigaciones Sociológicas and Ministerio de Agricultura, Pesca y Alimentación, Madrid, pp. 839–82.

Woolcock, M. (1998), 'Social Capital and Economic Development. Toward a Theoretical Synthesis and Policy Framework', *Theory and Society*, vol. 27, pp.151–208.

Chapter Nine

Regional and Rural Development in Austria and its Influence on Leadership and Local Power

Thomas Dax and Martin Hebertshuber

This chapter presents an overview of the situation of rural areas and an analysis of main factors of rural/regional development policy in Austria. Austria is characterised by a high degree of rurality and thus the 'rural' always has played an important role in the perception of society and policy. Problem regions in Austria, in general, have low population densities and therefore appear more or less rural in character. With the exception of some older industrial regions, the majority of the problem areas are fairly under-industrialised. However, agriculture is no longer the dominant economic activity in rural regions. The majority of areas have a multifaceted, although weakly structured sub-layer of SMEs, whilst tourism plays an important role in most parts of the Alps. Nevertheless, in terms of development policy, for a long time rural areas were hardly supported in an integrated sense or by territorial schemes, but primarily by sectoral measures. Only the concept of endogenous development introduced in specific remote parts of more mountainous areas provided a clear exception to the sectoral approach.

With accession to the European Union (EU), Austria's regional policy experienced a decisive break. Wide-ranging changes had to be dealt with across Austria's economic, social and political structure. These included the opening of markets (above all for agriculture), the introduction of new structures and supporting measures and the establishment of new institutional frameworks and policies in a short period of time. These changes also affected the rural areas and their economic and social positions.

This chapter starts with a short introduction of the position of rural areas within Austria's economic and social structures. The second part deals with relevant administrative structures of support for rural regions. Due to the complexity of these structures this can only be done partially and the paper focuses on 'bottom-up' movements in regional development. The third section explores major changes and impacts of the introduction and implementation of EU Structural Fund programmes. These new policy measures have influenced the former balances of leadership and local power and have triggered new dynamics in rural areas. However, it is too early to give more than a preliminary assessment of their actual effects. Nevertheless the conclusions attempt to outline both the main positive impacts and primary deficiencies of the rural development measures and institutions in Austria.

The Socioeconomic Structure of Rural Austria

Structure and Rural Areas of Austria

In terms of its location, Austria is often seen as lying in the heart of Europe. This geographical position had a strong impact on its post-war development which has been shaped by a rapid process of reconstruction. This development was influenced by two core factors, these being firstly the long period of border closure on the frontiers with the former East European countries which encircled nearly two thirds of Austria, and secondly by Austria's gradual integration within political and economic developments in western Europe. Despite its strong economic integration within western Europe, Austria has been characterised by a rather self-sufficient mode of development and its situation as an 'island'. Above all, the agrarian sector has been shaped to a great extent by specific national schemes preserving rural areas and in particular remote mountain areas, from international and global developments. The opening of the borders to the eastern countries in 1989 and the accession to the EU in 1995 have therefore been outstanding changes which have influenced the development of Austria much more than that of other EU countries. It is important to appreciate the social evolution of Austria and to interpret the resulting structural data against this background. The landscape and structure of Austria is shaped by a very high proportion of mountainous areas. Only in the eastern part is there a greater lowland area, with Vienna as capital and sole significant metropolitan area.

Different methods have been used to define rural areas but there has never been much consensus about this scientific and political discourse. To provide comparable data, we explain key features of rural areas by using the method adopted by the OECD's Rural Indicators Project (OECD, 1994). The first stage of this process requires the division into rural and urban communities at the local level. A community is labelled as rural if the population density is

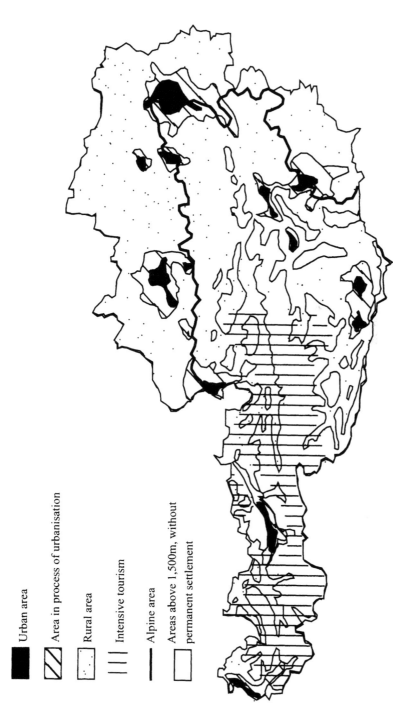

Figure 9.1 Elements of the large-scale spatial structure

Source: Österreichisches Institut für Raumplanung (ÖIR).

Urban area

Area in process of urbanisation

Rural area

Intensive tourism

Alpine area

Areas above 1,500m, without permanent settlement

under 150 inhabitants per km². In a second step a typology of regions is constructed with the share of the population living in rural communities calculated for each region. This share is used to divide the regions into three categories:

1   predominantly rural regions are those where more than 50 per cent of the population live in rural communities (often described as peripheral areas);
2   significantly rural regions are those where between 15 and 50 per cent of the population live in rural communities (also described as transitional regions);
3   predominantly urban regions are those where less than 15 per cent of the population live in rural communities (also described as economically integrated regions).

The OECD definition introduces a very different concept at the regional level comprising the centres and small towns of 'rural regions', in contrast to former definitions primarily based on the size of municipalities.

According to this spatial scheme, in all three categories there are both rural and urban communities. This hierarchical typology takes into account the intra- and inter-regional relations of exchange. In an international comparison of OECD countries, about 35 per cent of the population were defined as living in rural communities, which cover 96 per cent of the total surface (OECD, 1994). This share is particularly high in the North European and North American countries. This scheme also reveals the high share of rural areas in Austria. In Austria a high share of the population (42 per cent) lives in rural communities, whilst the EU average is 26 per cent. The geographical area of these communities amounts to 91 per cent of the total land area of Austria. However, in relation to the settlement area the rural areas are relatively densely populated. Pressure on settlements is very high, particularly in some inner-alpine valleys, those located inside the Alpine area with a limited proportion of land available for settlement. At a regional level, Austria together with the Scandinavian countries has one of the highest shares of rurality. According to the OECD classification, 78 per cent of the population live in the two types of rurally structured regions whereas the EU average is just 48 per cent (OECD, 1994).

Demographic and Economic Changes Affecting Rural Areas

The demographic development in predominantly rural regions of Austria is characterised by out-migration in most regions and continuing high birth rates in some regions. Therefore the population change remains slightly positive in most regions, although with a rising share of older people.

The second type, the significantly rural regions, are characterised by a very dynamic demographic evolution with population increases and a rising proportion of young people. This leads to significant tendencies of sub-urbanisation in 'transition' regions and high pressure on settlements and land in those communities. This trend has had an impact because of a rise in long-distance commuting relations between urban centres and rural regions all over Austria. This includes weekly commuting from remote (peripheral) regions such as from Waldviertel and South Burgenland to Vienna, and daily commuting from 'transitional' suburbanised areas.

Table 9.1   Socioeconomic indicators of rural and urban regions

Austria	PR[1]	SR[2]	PU[3]	Total
Area 1991(%)	71.4	27.5	1.03	100
Population 1991(%)	39.6	38.6	21.8	100
Population change 1981–91 (rates in per thousand p.a.)	2.8	4.5	1.2	3.1
Employment change 1981–91 (rates in per thousand p.a.)	1.7	7.2	3.6	4.4
Employment activity rate 1991 (% of persons aged 15–64)	65.4	65.4	67.1	65.8
Unemployment rate (%) of labour force 1991, national method	4.9	4.7	6.7	5.2
Employed women as share of total employment (1991)	38.8	41.3	45.4	41.2
Employment in sectors 1991				
Agriculture (sector 1)	13.3	4.1	0.8	6.2
Industry (sector 2)	37.3	36.9	29.4	35.0
Services (sector 3)	49.4	59.0	69.9	58.8

Notes

1   PR = predominantly rural regions.
2   SR = significantly rural regions.
3   PU = predominantly urban regions.

Source: OECD, 1994; authors' own calculations.

As in other rural regions the share of agricultural activities has decreased drastically during the last decade. In predominantly rural regions the share of employees in agriculture was 13 per cent (with rates up to 20 per cent in some remote regions) in 1991. A specific characteristic of Austria is the high rate of farmers with a second occupation in non-farming activities with an average of 60 per cent and peaks up to 80 per cent in some regions (South Burgenland, West Tyrol), due to the preservation of family farm holdings and the remarkable immobility of farm land. The very intensive relations between the agricultural sector and other economic sectors are expressed by this high rate of pluriactivity (Dax et al., 1995). In the trade and industry sectors there have been significant decreases in employment levels. It is worth noting that regional and economic policy has not taken this development into account, although this sector – specifically small industries, businesses, and handicrafts – has a high potential both for employment and innovation (Ax, 1997).

In consequence, employment in the service sector has increased steadily in rural areas, and above all in alpine regions with intensive tourism. In these regions the share of the service sector lies far above the average (more than 70 per cent). Regional policy measures have tended to be concentrated on tourist activities and in the past have favoured economic strategies supporting one particular sector (e.g. tourism in the Alpine regions).

These recent tendencies illustrate the differentiated nature of 'rural' regions which goes beyond a simple equation between rural and underdeveloped. The dynamics are very different between Austrian regions, and are basically influenced by tourism. The more disadvantaged regions have until now been mainly the border regions in the northern, eastern and southern parts of Austria. These differences in the economic development of rural areas are important both for the conception of regional analyses and the elaboration of strategies. As well as the spatial dimension, regions have been differentiated according to their development trends and potentials. This also means that a simple distinction between 'rural' and 'urban' is misleading and has to be complemented by a better understanding of the development dynamics of regions.

Definition of Rural Areas

In regional scientific and political discussions the term 'rural area' has often been used in a very undifferentiated manner as a synonym for all non-urban areas, or has generally been equated with peripheral (remote) regions. This cursory approach has been supported by methodological and conceptual problems of delimitation between rural and urban areas. Above all, the expanding 'transitional' space with fast changing spatial structures provides a wide field of interpretations. Thus the use of the term 'rural areas' was traditionally assigned to all those areas not being classified as 'urban', or to municipalities of a limited number of inhabitants (e.g. under 5,000 inhabitants: ÖIR, 1975, pp. 11–29).

However, the term 'rural area' often has been (and is) used without any concrete spatial definition and a common understanding of 'rural' is presumed. Rural space is commonly seen as those areas with agriculture as the dominant economic activity (and with landscapes obviously shaped by agriculture). The equation of agrarian policy with policy for rural areas could not be continued any longer because of progressive changes in most rural areas, including the decline of the economic and sociocultural significance of agriculture and the multifaceted nature of the interconnections between rural and urban space.

Because Austria has never had a specific rural policy, the need to define rural areas in Austria was limited. The different sectoral measures, also quite aware of the integrative manner of some of their programmes, focused on specific regional definitions or selection of pilot areas. With the application of EU regional policy, Austria had to find a way to decide for the first time a national area for rural policy. The categorisation of the Objective 5b regions (and of the Objective 1 regions) had to be done using the framework of the regional statistical units, drawing on NUTS-III level data, and had to be adjusted with the reference to other objective areas. The national preparation process for this regionalisation of rural policy provided a new incentive to commission research on these issues. The debate which ensued expanded the group of actors involved in discussions about regionalisation and to some extent prompted increasing concern for this issue. However, the adoption of the programming approach for the expenditure of finance under the Structural Funds lessened the importance of diversity between regions and supported targeting of assistance for each type of objective area, thereby weakening the regional approach (Dax, 1995).

Administrative Context and the History of Development Policy in Austria

Institutional Framework of Regional/Rural Development

The administrative structure of Austria is characterised by a division of the state into nine provinces (Länder), with a comparatively high autonomy in legislative matters and policy implementation, and some financial autonomy. Districts, a territorial unit between Länder and municipalities, have predominantly administrative tasks. Control over regional policy and local development is not held by a single public institution but rather is allocated to various institutions on three territorial levels. According to the federal constitution the municipalities are responsible for local development planning. Responsibility for the regulation of all other aspects of territorial planning lies with the Länder. The federal state can only take decisions relating to planning issues in the fields in which it has competence, (such as trade and industry, water and forests, and mining). The federal state also has responsibility for the coordination of territorial policies. With Austria's accession to the EU, a fourth, supra-national level has been added.

Because of the small scale of the Austrian economy, and because of Austria's sensitive landscape, tasks concerning regional development policy cannot be separated in terms of territory or sector. The wide distribution of competencies in regional development policy is an obstacle to the construction of linkages between development planning and regional policy. In addition, both tasks are cross-sectional and in a weak position compared with 'strong' sectors and their respective administrative departments such as agriculture, trade and other traditional branches of the rural economy.

As well as legislative development planning, there are many non- or partly governmentally regulated activities of great relevance for regional policy. Generally, infrastructure and services are allocated to regional administrative bodies. Funds and information services about regional development are jointly carried out by the federal state, the Länder and the municipalities. For this reason, the various public institutions and levels of government often act simultaneously both in parallel and in competition with each other, as well as on a more informal cooperative basis (ÖROK, 1996, p. 51). Since 1971 the federal government, provinces and local authorities have promoted the Austrian Conference on Spatial Planning (ÖROK, Österreichische Raumordnungs-konferenz) as a common advisory platform in which social partners are also represented. The ÖROK plays a key role in coordinating regional policies within a wider national framework under development for the coordination of spatial economic and regional policy.

These spatial planning activities are the main driving force behind Austrian discourses of regional development. However, they remain in general noncommittal. The ÖROK engages with important issues of social development and spatial planning, and formulates objectives for recognition by the regional and local authorities. The federal state can raise 'consciousness' about spatial planning issues, but cannot decide and impose obligatory measures (Fassmann, 1997, p. 4).

The ÖROK has paid special attention to the spatial development of peripheral regions and to the formulation of integrated development policies. An example would be the consideration given to the contribution of mountain farmers to the maintenance of the cultural landscape. In the Austrian spatial planning strategy of 1991 (ÖROK, 1992), which enumerates problems, objectives and measures of important issues of spatial relevance, the maintenance of the cultural landscape as a framework for recreation and tourism, the avoidance of erosion, and the role of the forests for recreation and prevention of natural hazards are emphasised as crucial functions of ecological and sustainable agriculture. Various experiences of existing regional policy could provide a basis for the application of the Austrian structural fund programmes.

Regional/Rural Policy Approaches in Austria

The economic prosperity of the 1960s and the 1970s facilitated a programme of comprehensive public expenditure which aimed at reducing a substantial and historical backlog of demands for infrastructure and services in rural areas, such as transport, communications, education and health facilities. This was achieved primarily by making funds available for establishing enterprises to create employment for those made redundant by the shrinkage of the agricultural sector. In combination with the development of tourism in the core regions of alpine tourism, this led to the assumption that a decrease in regional disparities would be generally feasible.

Therefore until the beginning of the 1980s a specific focus of 'top down' regional policy was the encouragement of growth of economically weak regions through external support. Until the 1970s this policy was based on (and funded by) significant economic growth in the centres of industry. Regional policy was conceived of as the spatial reallocation of resources generated by this growth. In fact, a decisive contribution to the diminution of regional disparities was achieved, but this process was matched by the neglect of the development potential inside individual regions.

Due to the persistent pattern of regional problems, regional policies have attained high political importance. Particularly since the 1970s, the broad understanding that something had to be done for peripheral areas has led to a number of cooperative measures at different institutional and territorial levels.

In the following section we present an overview of the evolution of 'bottom up' or endogenous development initiatives, their specific theoretical base and experiences.

From 'Independent' to 'Endogenous' Regional Development

With the economic crisis and the reduction in levels of economic growth in the 1970s, it became evident that regional disparities could hardly be reduced by conventional means under the new structural conditions of growing unemployment and so on. By the end of the 1970s, Austria's economic situation was characterised by an increasing polarisation between centres and peripheral, economically weak regions caused by an economic downturn and the emerging problems of structurally weak older industrial areas. On the other hand, the ecological questions concerning the consequences of progress and prosperity in industrial societies led to a growing interest in questions of ecology. In addition, the rural, provincial areas began to be re-evaluated in sociocultural terms, with the disappearance of local and traditional ways of living and economy interpreted as cultural loss.

The increasing criticism of the objectives and measures of agricultural policy was supported by diagnostic studies (Krammer and Scheer, 1980) which clearly foresaw the future difficulties of agriculture, in particular the consequences of surpluses of agricultural production. The problems became particularly serious in those rural areas where a number of conditions coincided. These included aggravated conditions for production in agriculture; a lack of job possibilities in trade, industry and the tourist sector; long distances to work centres (including weekly commuting); and lack of attractiveness for tourism (Bundeskanzleramt, 1981, p. 9).

In some specific peripheral regions (e.g. Waldviertel and Mühlviertel) the first experiments in 'alternative' regional development were started. Initiated by political discussions about theories of development (such as centre-periphery models and dependency theory) active groups emerged with demands for autonomous or independent ways of development. Existing disparities between regions were made more visible by various activities, for example, through the sale of potatoes from the Waldviertel to consumers in Vienna at the farmer's price. The first of these self-help activities was the foundation of producer-consumer associations by young farmers and consumers in the city. These associations later provided a model for a new conceptualisation of independent regional development.

The motor for the development of the first policy instruments was the Austrian Mountain Farmers Association (ÖBV, Österreichische Bergbauern-vereinigung), founded in 1975 by young farmers. To establish an organisational basis for the promotion of new development approaches, a specific fund was created in order to draw on the knowledge of experts and to provide support for new project initiatives.

In 1983 the Austrian Association for Independent Regional Development (ÖAR, Österreichische Arbeitsgemeinschaft für Eigenständige Regional-entwicklung) was founded, based on an alliance of regional consultants and regional associations. This served as a platform in the regions for the exchange of ideas and the promotion of sociocultural activities. From then on until Austria's accession to the EU in 1995, the ÖAR – commissioned by the Federal Chancellery – established regional consultancy structures in order to provide regional support. In 1985 the fund was reorganised, becoming the Aid Initiative for Independent Regional Development (FER, Förderungsaktion eigenständige Regionalentwicklung).

In this pioneering phase, the model of independent regional development was based on three pillars:

1 regional associations of pioneers and local actors, creating and discussing initiatives and developing local projects;
2 regional consultancy activating local people and providing professional support for innovative projects;
3 support funds providing free consultancy and subsidies for establishing innovative and cooperative projects particularly in the phase of project conceptualisation and project initiation.

In the middle of the 1980s, associations for the promotion of endogenous development already existed in nine small regions, mainly carried out by core groups of local activists.

We would argue that this approach meant a fundamental change in the paradigm of regional development, from a 'catching-up' modernisation to 'bottom-up' (independent, innovative, endogenous) approaches towards regional development. Moreover, increasing public criticism of the destructive effects on the environment and the increasing disparities caused by modernisation widened the discussion of what regional development should include within its objectives. This shift in development policy was partly accompanied by theoretical work. However, regional development discourse in Austria has been shaped primarily through practical experiences and has rested on a theoretical base that has been narrow in terms of both scope and quantity (Dax, 1995). The new paradigm can best be described as that of the weak region no longer solely an object of 'top down' (federal) regional policy and a self-governed subject from 'bottom up'.

This new orientation was formulated in the concept of Independent Regional Development (ERE, Eigenständige Regionalentwicklung). In this, notion of a standardisation of all living conditions was questioned, and the revaluation of rural structures and the importance of internal regional potential and resources was emphasised. Development was no longer seen as problem-solving on behalf of the people of a region, but rather as the discovery of solutions by the people of the area themselves.

ERE, in contrast to sectoral and functional strategies of development was characterised by an area-based sociopolitical concept. From the beginning the focus lay with the implementation of sustainable and environmentally sound projects to improve the structure and economy of the region. The objectives of ERE were:

1   to foster regional identity, motivation and the initiative of the population;
2   to increase regional added-value, based on a region's own resources and potential;
3   to construct a multipurpose and stable economic structure, based on region-specific quality products; and
4   to achieve compatibility between economic measures and the cultural and ecological characteristics of the region (Bundeskanzleramt, 1981, p.31f).

Projects had to be innovative and experimental in character, self-determined in terms of the organisation of regional cooperation, and orientated towards the creation of democratic forms of organisation and employment, promoting further education and self-reliance. They also had to be multisectoral in terms of their cooperation.

As a result of the experiences of the first projects, it was emphasised that for the creation of innovative projects a climate of learning and activation, a long-term phase of creation and intensive discussions and exchange of experiences might be necessary. This climate could only be created by specific and organised work in cultural and further education. The approach of self-reliance was guided by methods of community development as a strategy for enabling the active involvement of the population. The barrier of 'structural power' could only be broken through by critical and participatory work.

It soon became clear that the standards demanded by regional consultants were too high. Above all they found themselves more and more in a dilemma between professional consulting on economic projects and the more general demands of the regional societies. Therefore from 1985 onwards the work of the consultants was restricted to the provision of information and counselling to economic projects. Cultural and political activities were excluded from their tasks and action research was abandoned. Thus after 10 years of independent regional development a new structure and increasing professionalism of the ÖAR was seen as necessary in order to respond to the experiences and changing needs of regional development. Scheer (1990) includes the following as causes of these changes:

1 the economic structure was not effective enough (at least in the short term);
2 the regional consultants were confronted with contradictory expectations by radical and regional exponents of the regional associations and the success-oriented subsidy provider;
3 the acceptance of the concept behind ÖAR in the regions was insufficient. Some businesses and decision-makers could not identify with the aims and methods of ÖAR.

The shift in the organisational concept was intended to promote a clear image of the profession and of efficient organisational structures. ÖAR understood itself as a dynamic regional consultancy agency with advanced expertise in economic issues. Consultancy activities have been extended continuously and have included regional analysis, the formulation of development programs, and consultancy for municipalities and local officials, economic processes and projects. The ÖAR was transformed into a Limited Liability Company (GmbH), with the new focus of occupation on services offered on the basis of contracts.

At the beginning of the 1990s a more innovation-oriented strategy gained in importance. Innovation and strategies for the adaptation of technology are key factors to the successful improvement in the development of enterprises and regions (ÖAR, 1994; cf. also Danielzyk, 1998). The marketable and innovation-oriented regional policy aims to recover regional competition, in the frame of the international division of labour and markets, to enable the regions to succeed in the 'competition of the regions'. The movement is currently towards policies to overcome sectoralisation within regional structures and to create horizontal and integrated networks.

The 'Leitbild' (the main priority and perspective of a region) of endogenous renewal is a multifaceted, flexible and dynamic economic structure, within which enterprises can compensate for their handicaps in size and location through intelligent forms of cooperation. Through flexible production for market niches and high claims on quality they should be able to obtain self-reliant growth and a higher quality of life (ÖAR, 1994, p. 6).

The main criteria and strategies for success of endogenous regional development were the exploitation and integration of regional potential and strengths; the active inclusion of as many local and regional actors and decision-makers as possible; the development of a broadly accepted 'Leitbild' and its objectives; the handling of the region as spatial entity of location, exploitation, decision-making and identity; and an understanding of the region as network-system with its internal and external interactions.

In conclusion, it appears that the concept of independent regional development and the corresponding regional policy approaches has undoubtedly had an essential influence on the further conceptual development of Austrian regional policy.

Ultimately, the concept of ERE can look back on a very positive record in recognition of its achievements at national and international levels. By splitting or creating new initiatives, additional facilities and institutions such as cultural, employment and environmental consultancies were established to support regional initiatives. The ideas of ERE have diffused into different branches and have been included into other, partly competing programmes at a Länder-level, village renewal and regional management would be examples. At the beginning of the 1990s the discussion and development of concepts in science and politics was fully under way. Endogenous development was recognised in an international context and experiences have been transferred to other regions.

Despite these successes it should be recognised that these policy approaches for disadvantaged regions have not been sufficient to compensate for the disadvantages of their geographical location. At best – when a certain regional dynamic arises from individual projects as a result of the multiplier effect and successor projects – regional policy has succeeded in exploiting new development potential and halting the downward economic and social spiral (Hovorka, 1998, p. 45).

In the first half of the 1990s the European integration process and dramatic changes in central and eastern European countries were the main forces which influenced the regional development and policy in Austria. The new situation in the larger territorial context and its impact on regional development had to be taken into account in preparation for Austria's participation in EU regional policy. The elaboration of the 'regional development concepts' in particular had to reflect recent changes in location factors. The debate primarily focusing on these two changes has neglected to some extent the influence of globalisation processes, independent from the Central European situation, but also extremely relevant on economic and social trends in Austria.

Those development concepts, commissioned by the federal state and the Länder, have not only supplied a rather complete coverage of analysis and development concepts on the situation of most of the rural areas (and also other problem areas) in Austria, but have stimulated at the same time the discussion of regional/rural development issues and provided a basis for rural development programmes (the Objective 1, 2 and 5b programmes and community initiatives such as LEADER II, INTERREG II and SME).

EU Regional Policy in Austria: Implementation and Experiences

Application of EU Regional Policy

In Austria regional policy – developed relatively independently by the federal state and the Länder until 1990 – experienced a massive change with Austria's accession to the EU. The first change concerned the increase in public interest in regional policy, which developed during Austria's negotiations for EU membership.

With the accession to the EU, Austrian regional policy came under pressure. On the one hand, regional funding represented a possibility for the return of funds from the EU as compensation for membership payments, which were viewed with suspicion by some people. On the other hand, because of the new system of regional policy coming from outside, EU accession provided an opportunity to question the hardened administrative structures in Austria (Fassmann, 1997). Also, politically, this opportunity provided a suitable occasion to break the widespread scepticism held in rural areas about the consequences of EU-accession.

The new formulation of regional policy in Austria had to find appropriate organisational structures for the implementation of the EU programmes which had to be build up relatively swiftly. The great time lag between programme discussion, approval and implementation led to considerable disappointment and the focus of regional policy shifted more to the aspects of funding.

To a great extent, experiences with previous approaches of innovation-oriented regional policy have been picked up with regard to the implementation of the EU programmes (especially Objective 5b and LEADER II). Due to the wide application of regional policy a host of new experiences have been encountered. From the start regional management agencies have been built up through cooperation between the Federal Chancellery, Länder (provinces) and municipalities. These have contributed to successful implementation. The objective-areas of EU structural funds (cf. Figure 9.1) comprise the essential problem areas and underline the 'rural' character of wide parts of Austria (cf. Table 9.2).

Table 9.2   Characteristics of Objective 5b areas in Austria

	Carinthia	Lower Austria	Upper Austria	Salzburg	Styria	Tyrol	Vorarl- berg
Surface (km²)	8.365	12.548	8.163	4.073	8.159	7.766	1.000
Population 1991	327.264	617.912	545.663	88.644	472.203	190.607	39.500
Population density 1991 (EW/km²)	39	49	67	22	58	25	40
Pop. development 1981–91 (in %)	+0.8	-0.8	+5.1	+4.3	+2.1	+6.5	+9.8
GRP/Inh. 1989 (EU=100)[1]	68	69	61	87	70	83	_[2]

Notes

1   Gross regional product per inhabitant (EU=100), calculated on regional level (district or NUTS III).
2   Not available.

Source: Dax and Oedl-Wieser, 1995; Dax, 1997.

The Objective 5b areas covered 60 per cent of the total surface of Austria and comprised 29.2 per cent of the population. This percentage is the highest in all the EU-member states (France is 17.3 per cent and the EU average is 8.2 per cent) and thus illustrates the importance of rural areas in Austria. Only a small part (Burgenland) is classified as an Objective 1 area. The differences between the Objective 5b areas are clear, for example in the development of the population, which has increased in the western (mountain) areas due to the high birthrate, whereas in the other areas it stagnates. The marked regional differences are clearly evident in the figures for gross regional product per inhabitant, which also falls from west to east. These regional differences are primarily caused by the development of tourism.

The high rate of participation in the LEADER II programme (32 small regions), in which especially innovative programmes and measures are supported, should also be pointed out. The main work of the local action groups (LAGs) is supported by a professional 'Leader-manager'. Since the accession to the EU, in nearly all regions 'bottom up' oriented regional support structures have been established for the development of Objective 5b areas.

Experiences with EU Regional Programmes in Austria

The concept of endogenous regional development was used in part in the development of guidelines for the implementation of the regional economic development programmes. However, because of the long period between the drawing up of the concepts and the start of the programmes there was a reduction in willingness to develop projects. These problems with the initiation of projects had rather unfavourable effects when the implementation of the programmes started and made it more difficult to realise innovative ideas. The installation of the regional managers in some cases took place very late and therefore could not work against the loss of motivation in the regions. Furthermore their role was not clarified and was overloaded with high expectations (Dax, 1997).

When the 'dust of the whirl caused by the conceptual and organisational initial stage' eased off (Schremmer, 1997, p. 18), some of the deficiencies and problems in the implementation of the new programmes and structures in regional policy became more distinct. In the past few years, regional policy and regional development in Austria has changed very fast from an initial phase of new stimulation and new structures to a phase of overstretching and lack of orientation. According to Schremmer (1997) these changes have been caused by over-optimistic expectations prior to Austria's accession to the EU; insufficient reflection about their own position, needs and connected requirements amongst politicians and activists in regional policy; disappointment when expectations have not been reached; and the realisation that the requirements to achieve positive results are much greater than had been imagined. There were also fundamental misunderstandings with regard to Austria's experiences in regional policy. The assumption that Austria had wide experience and practice of many years with the conception and implementation of regional policy turned out to be over-optimistic.

Ultimately, experiences with the implementation of regional development programmes and policies have limited people's expectations. According to some researchers, a substantial reason for this was that at a regional level the institutional and organisational prerequisites were not sufficiently established to develop and implement endogenous strategies. In contrast to the self-image

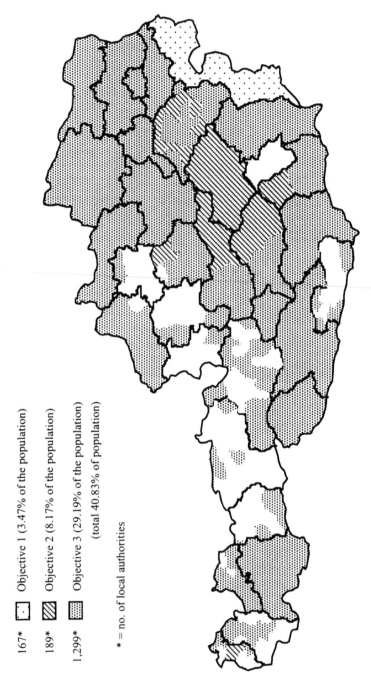

Objective 1 (3.47% of the population)   167*

Objective 2 (8.17% of the population)   189*

Objective 3 (29.19% of the population)   1,299*

(total 40.83% of population)

* = no. of local authorities

Figure 9.2 Objective areas under the EU Structural Funds in Austria

Source: ÖROK (Österreichische Raumordnungskonferenz), graphic ("treatment") by
M.Kogler, Bundesanstalt für Bergbauernfragen, Vienna 1997.

of Austria as a country of innovation in regional policy, many regions had no structures and insufficient experiences with 'bottom-up' approaches in regional policy.

A large part of the sectoral support and support structures introduced before Austria's EU-accession continued to exist and were linked to the Structural Funds and programmes. These are structures such as the regional bonus for innovation, regional infrastructure support, and procedures for establishing and enlarging centres of technology transfer and innovation. In addition, the FER (support activity for independent regional development) was brought into line with the tasks and the realisation of the EU regional programmes, in particular for setting up regional management and regional concepts.

The implementation of the EU Structural Funds and community initiatives had to undergo an interim evaluation. The main results of the midterm evaluation were summarised and formed consciously trenchant statements for the Objective 5b and Leader II programmes.

There were four key findings (ÖROK, 1999). First, the Austrian Objective 5b programmes for regional development are characterised by a strong sector dominance at both the planning and implementation stages. There is a clear division in the process of operative implementation in the fields of agriculture and forestry, economic support of the production and service industries, and the labour market. This division parallels the administrative demands of the EU to separate the Structural Funds into three strands: EAGGF, ERDF and ESF.

Second, as regards their content, the main emphasis in the implementation of the programmes is on material measures (like infrastructure projects and the fostering of material investments). The concentration on material investment and infrastructure measures remains typical even from an international point of view and – for various reasons – is to be regarded as a 'usual' approach for regional development: it is easier to 'sell' politically, it allows a quicker financial return on investment and can be documented easily in monitoring. Compared to that, support for 'soft' factors such as capacity building and further education is only beginning to develop.

Third, the idea of including the regional level in the realisation of programmes has not really been translated into action, and at the regional level the sectoral orientation dominates over an integrated strategy for (regional) development.

Fourth, the potentials for synergy between the three Structural Funds are scarcely used, as a result of poor coordination between regional agencies. The use of synergies between various support activities would require the development of an information exchange and cooperation system between different administrative units which would take a longer period of time. Attempts to completely restructure administrative structures to ensure synergies can cause resistance, limiting the extent to which aims can be achieved.

The programme of work set out in the Single Programming Documents is carried out by different administrative representatives of the federal state and Länder thus taking advantage of existing administrative structures. Existing knowledge about the sectoral handling of administration of supported projects can be used most efficiently for the operation of the Objective 5b programs. New institutions are only set up where adequate administrative structures do not exist, for example accompanying committees of representatives of the EU, federal state, Länder and social partners (which include trades unions and industry pursuing a policy of consent) for sectorally overlapping coordination and evaluation of programmes. This also means that the initiation of projects is characterised by a clear sectoral division which makes the adoption of integrated regional support difficult. Additionally, the membership of the committees favours the usual paternal structure of the clients of these supporting institutions and the individual and sector character of the support. This is a disadvantage for innovative projects.

The operation of the LEADER II programmes, which are conceptually close to the original intentions of the Austrian ERE concept, is carried out to a large degree within existing administrative structures. At Länder level coordination and administration agencies had already been established. Further support structures have been added, in special departments of the Länder government through outsourcing to private support agencies. In the federal state the Federal Chancellery coordinated programmes and measures. Austria was the first country to establish its national LEADER network. At a regional level the local action groups (LAGs) were usually constituted as registered societies and were responsible for carrying out programmes on the spot (submission of project plans, coordination of local initiatives and participants).

The interim evaluation report has examined the small structure of the Austrian LEADER areas and points to deficiencies in the institutional framework and networking activities. It concludes that no explicit measures in the field of equal treatment of women and men can be identified. The domination of men in the programme execution should not be interpreted as a specific trait of LEADER II programmes but might rather mirror traditional forms of allocating responsibilities to men and women in rural areas.

Installation and Role of Intermediate Regional Management Agencies

For initiating projects at a regional level so-called regional management centres were installed in most of the Objective 5b programme areas. In these, there is a better chance of the development of comprehensive and integrative projects (ÖROK, 1999, p. 69).

We have to distinguish between two types of regions. Some regions have specific (if very different) experiences with regional policy. Other regions have less experience, and structures for regional policy can only be used indirectly and mediated by the consultancy agencies which helped to write the programme documentation and develop administrative structures. Moreover, the need for regional management agencies as the central model of organisation for the implementation of EU regional policy was anticipated by programme coordinators at the federal state level.

The Federal Chancellery had suggested that regional management should become the institutional base for the EU programmes. From 1996, with time pressures, new requests for EU policy and high expectations, 25 new regional management agencies were established. Regional management was understood as 'multi-functional interface, turntable and networking, which informs and connects the local and regional actors and therefore cooperates with Bundesländer, national and EU-institutions' (Scheer et al., 1998, pp. 22ff). The key tasks of regional management agencies are the provision of advice and information, consulting for projects, the development of projects and initiating improvements in networks of local actors and institutions (Scheer et al., 1998).

An interim evaluation of the work of regional management in Austria (Scheer et al., 1998) showed that most of the regional management agencies are organised as associations of municipalities with membership usually comprising all the municipalities of the area. In addition the social partners and the elected representatives of the region (members of the Landtag, Upper House, Federal parliament and the EU parliament) are also represented. It is evident that the new institutions are built on existing administrative structures almost without exception, and that in many regions attempts have been made to include political parties and social partnership. In Austria, the 'social partnership' constitutes a political and rather influential forum for representatives from the main interest groups (economy, industry, workers unions and agriculture). However, there have been questions about procedures to deal with any absence of consensus in aims and priorities, and problems with the integration of innovative persons into institutions dominated by representatives from politics and administration. There are questions still about the development and appropriate presentation of regional strengths.

Among regional managers there is a general agreement that regional management is and should be a more regionally oriented service, near the region and carrying the region. It also needs a variety of orientations, structural flexibility and openness in culture to be able to fulfil the core function of horizontal and vertical networking.

Regional development agencies need to be based on appropriate structures in the region. In many regions there are still questions about the position and functions of regional management. However, the two main tasks of regional management are undisputed. The first is the development of projects. 'Project development is to be understood as the process of activating and forming up by which potentials and ideas become projects' (Scheer et al., 1998, p. 36). As experience shows, this phase is usually the bottleneck of regional development. Complaints have been made that there are enough (or even too many) well justified projects and suggestions for strategies. However, most include rather vague concepts with regard to the use of regional strengths. There are also shortages in personnel to carry out the projects. Regional management has the task of managing the process by which people and structures can be prepared for carrying out successful projects. The second task is the networking of regional actors. 'Regional management is developing networks, managing networks, bringing together actors from different regional, professional and functional contexts, and mobilising and securing resources in the network' (Scheer et al., 1998, p. 37). A decisive task of regional development therefore is to create structures which can use competencies and the resources to connect existing institutions. Regional management agencies therefore should be measured by the number and quality of their networking and not primarily by projects completed.

Crucial themes for the future of policy development include consideration of the regional management and the Länder. Länder are partners, financiers or owners of the regional management agencies. Therefore the quality of relationships between regional management and each Land is central. There seems to be a consensus that regional management is an intermediate task and depends on intensive and effective exchange of expectations and services between regional management and Land.

Regional management agencies need access to and the involvement of local activists as well as easy access to the relevant institutions of the Land. Both trust and 'cultural affiliation' are important factors of success. At the same time the capability to look at structures and projects from a critical distance is expected. This results in an intermediate position for regional management within a highly complex system. Useful measures to prevent failure would be the use of contracts, independent financing and strategic partnerships.

Consideration of the tasks and services of regional management in terms of client relationship will also be a theme for the future. There is a consensus about the core tasks of regional management, but the pressure to achieve immediate success remains (Scheer et al., 1998, pp. 45ff). Regional managers have to ensure that the funds from EU programmes will be used in full. Regional managers have to set up key regional projects. Regional managers have to tackle long-standing development problems.

A central question concerns the conditions under which regional management agencies should become carriers of projects. There is an indication that projects are prepared, but after that carriers for their implementation are often missing. This lack of initiative and entrepreneurial activity is more marked in structurally weak regions. Regional management should engage in projects only if they comprise regional and complex subjects related to the general infrastructure of the locality and region.

There are also issues surrounding the relationships between regional management and the region. The intensity and quality of the relationship with the region is described as the most important strategic component of regional management. Because of the great complexity in the network of relations, regional management needs a core of regional decision-makers and multipliers, which take over an active function of ownership. Otherwise regional management would quickly lose anchoring and weight.

An investigation of the roles of local experts in regional policy (Bratl et al., 1998) found that the new regional management agencies had already achieved a high and widespread degree of recognition and effectiveness. They are the first point of contact for information for municipalities and project-leaders. This monopoly of existing administrative and associative structures in information provision becomes evident at the project level. The (planned, submitted or completed) projects are distributed as follows: 35 per cent to tourism, 24 per cent to agriculture, 15 per cent to qualification, 14 per cent to productive trade, 7 per cent to culture, 7 per cent to industry, 6 per cent to energy and 6 per cent to multi-sectoral regional development.

Conclusions

The early preoccupation with endogenous regional development in Austria has promoted the debate on regional and rural policy. After a first euphoric phase of bottom-up development which initiated a lot of innovative projects, it became clear that the structures for promoting such development could not be established without or against traditional political and administrative structures. However, the pilot programmes led to intensive discussions about regional development and showed enthusiasm for the innovative and alternative actions associated with an endogenous approach.

Because of this, decision-makers had to offer new approaches to development. A period with very different approaches and experiences followed. Subsequently actions like village renewal, direct marketing, organic farming etc. became standard in local development. The process of professionalisation intensified and the idea of independence in development was replaced by innovation, and the more individual bottom-up approach was replaced by a systemic view on regions and their development.

'Networking' is now the key word and principle of regional development. This works from the assumption that networks can be installed in the regions if commitment by decision-makers and local actors to cooperate can be achieved. Although the capacity of regional initiatives for action were partly overestimated, a common understanding was reached of the necessity of intra- and inter-regional cooperation as a central challenge of regional development.

Austria's accession to the EU meant a fundamental change in the practice of regional policy. The extension of regional policy led to new dynamics and ideas in many regions. The realisation of the need for the participation of wider social groups and intermediate networking advisory agencies in order for an integrated approach to regional development for it to work properly led to the general installation of regional management agencies.

However, the traditional structures remained in place and received new responsibilities which they fulfil by adapting only partially to the new requirements. The division of agendas and the sectoral and hierarchical organisation of policies and administration has proven to be very resistant. Strategies for change have confronted the political and social reality of regional contexts. Thus we assume that innovative ideas and initiatives failed due to structural conditions in political and administrative systems.

There is no general agreement on the merits (and shortcomings) of evaluation in Austria and therefore the recent experiences of regional policy have not yet been evaluated. On one hand, successful implementation of the programmes shows the importance of integration and networking strategies. On the other hand, failures and deficits are often due to the strong resistance of sectoral and partial interests and views, above all in regions with no long-term experiences with the planning and implementation of regional development strategies.

In conclusion, the strengths and weaknesses of the Austrian approach to regional development are as follows. First, from the end of the 1970s and beginning of the 1980s efforts have focused on the least developed (peripheral, in most cases small) areas. With the initiation of pilot projects and consulting as a central point of project development the approach of independent/ endogenous regional development was fostered and showed promising results. This success of these pioneers has given rise to an interesting discussion on the issue and led to a further refinement on the theory of endogenous local/ regional development.

Second, the federal administrative structure and the far-reaching distribution of responsibilities for regional/rural support to a number of different institutions both at national and provincial level resulted in a great number of different measures and a support pattern which cannot be analysed easily. This support structure led to some difficulties, but its flexibility allowed specific problems to be addressed and can be seen as one of the reasons for the relative success of Austria's regional development. 'The positive impacts of untidiness' are also seen to be one of the essential contextual features to allow the emergence of innovation in rural (peripheral) regions (OECD, 1995).

Third, debates about regional/rural development received a main impetus from the core aim to reduce territorial disparities between urban and (peripheral) rural areas. But the territorial dimension of the rural aspect is not clearly worked out in many research projects in Austria. Implications of definitions are neglected by the tendency to remain conceptually with a rural-urban dichotomy model although this has proved to be rather outdated in many respects (cf. Holzinger, 1995).

Fourth, rural is still often confused with agricultural issues in general. 'Rurality' is politically dominated by the concept of 'rusticality' (*Bäuerlichkeit*), which is closely connected with conservative ideas and attitudes. Research concerning rurality is to a great extent dominated by this conservative view.

Finally, there is a lack of research into the impacts of regional development on the social and political structure in affected regions, the role of actors and institutions, and the social groups being involved and profiting from regional development. (An exception is the research from Tobanelli (1997) about the implementation of EU regional policy in Salzburg.) There is every indication that – with the accession to the EU – the 'centre of gravity' (and of profits) of regional/rural development has shifted from the fringe groups to the (local) elites.

References

Ax, Ch. (1997), *Das Handwerk der Zukunft. Leitbilder für nachhaltiges Wirtschaften*, Birkhäuser, Basel.

Bratl, H., Fidlschuster, L., Weinberger, Ch. and Mitarbeiter (1998), *EU-Förderungen für Regionen. Informationsstand, Informationsbedürfnisse und Zufriedenheit regionaler Entwicklungsträger*, Invent-Institut für regionale Innovationen, Vienna.

Bundeskanzleramt (1981), *Eigenständige Regionalentwicklung. Ein Weg für Strukturell Benachteiligte Gebiete in Österreich*, Bundeskanzleramt, Sektion IV/Abt.6, Raumplanung für Österreich 1/81, Vienna.

Danielzyk, R. (1998), *Zur Neuorientierung der Regionalforschung*, Wahrnehmungs-geographische Studien zur Regionalentwicklung 17, University of Oldenburg.

Dax, T. (1995), *Research on Rural Development: Review of the Situation in Austria*, Paper to the first REAPER Seminar, Stadtschlaining, September 1995, Vienna.

Dax, T. (1997), 'Ziel 5b: Aufwertung des ländlichen Raumes', *RAUM, Österreichische Zeitschrift für Raumplanung und Regionalpolitik*, vol. 25, pp. 26–8.

Dax, T., Loibl, E. and Oedl-Wieser, Th. (eds) (1995), *Pluriactivity and Rural Development, Theoretical Framework*, Bundesanstalt für Bergbauernfragen, Research Report No. 34, Vienna.

Dax, T. and Oedl-Wieser, Th. (1995), *Ex-ante Evaluation of the Rural Development Programmes (1995–1999) for the Objective 5b Regions in Austria*, Evaluation report for the Directorate-General for Agriculture (DG VI) of the Commission of the European Communities, Vienna.

Fassmann, H. (1997), *Raumordnung, Raumplanung und Regionalpolitik in Österreich*, GWU-Materialen 2/97, Vienna.

Holzinger, E. (ed.) (1995), *Rurbanisierung: Abschied von Stadt und Land*, Österreichisches Institut für Raumplanung, Vienna.

Hovorka, G. (1998), *Rural Amenity in Austria. A Case Study of Cultural Landscape*, OECD, Vienna and Paris.

Krammer, J. and Scheer. G. (1980), *Das Österreichische Agrarsystem*, Bundesanstalt für Bergbauernfragen, 2 volumes, Vienna.

ÖAR (ed.) (1994), *Handbuch Ausbildungsprogramm Regionalberatung*, ÖAR-Regionalberatung, Vienna.

OECD (1994), *Creating Rural Indicators for Shaping Territorial Policy*, OECD, Paris.

OECD (1995), *Local Responses to Industrial Restructuring in Austria*, Country Report on Austria, OECD, Paris.

ÖIR (Österreichisches Institut für Raumplanung) (1975), *Der Ländliche Raum in Österreich*, ÖIR-Veröffentlichung 37, Vienna.

ÖROK (Österreichische Raumordnungskonferenz) (1992), *Österreichisches Raumordnungs-konzept 1991*, ÖROK Schriftenreihe No. 96, Vienna.

ÖROK (1996), *Position Österreichs im Rahmen der Europäischen Raumentwicklungspolitik*, Österreichische Raumordnungskonferenz, Vienna.

ÖROK (1999), *Zwischenbewertung der Ziel 5b- und LEADER II-Programme 1995–1999 in Österreich*, ÖROK Schriftenreihe No. 144, Vienna.

Scheer, G., Baumfeld, L. and Bratl, H. (1998), *Regionalmanagement in Österreich. Eine Zwischenbilanz*, Invent- Institut für regionale Innovationen, vol. 3, Vienna.

Schremmer, Ch. (1997), 'Noch vor dem Start am Ende? RAUM', *Österreichische Zeitschrift für Raumplanung und Regionalpolitik*, vol. 25, pp. 18f.

Tobanelli, T. (1997), *Regionalförderung als Politikum*, University of Salzburg, Salzburg.

Chapter Ten

Recent Rural Restructuring and Rural Policy in Finland

Torsti Hyyryläinen and Eero Uusitalo

Rural Finland on the Watershed

Finland is a place with a character all of its own. It is rural by its nature, with vast forests, lakes and rivers, but its towns are tiny by European standards, and it is only the capital, Helsinki, with its surrounding areas that could be called a metropolis in any way with its population of 1 million. Finland is the most sparsely populated country of the EU (15 people per km^2), with a total population of 5.1 million and a land area of 336,594 km^2. Only 20–25 per cent of the Finns live in rural areas (about 1 million). Finland is a place where East and West merge. The history, culture and religion of the country place the Finns squarely among the peoples of Western Europe, and yet there is a strong eastern influence.

The period 1989–99 marked a significant watershed in the history of Finland, a time when old structures were demolished and new ones created. The drastic opening up of the economy to international influences in the course of the 1990s forced the Finns to accept new ways of doing things – and brought them face to face with the severe challenges of internationalisation. The upheavals that took place in Russia, and EU accession in January 1995, pointed the way to changes that affected all aspects of society.

Nevertheless, a strong local element is a crucial factor in rural development, which is grounded in the country's history, in a civil society that functions on the strength of local democratic organisations, municipal authorities and about 190,000 voluntary organisations of one kind or another. The country's small farms, small businesses, citizens' associations and small local government units are actively searching for new survival strategies and thinking more seriously than ever about the benefits of cooperation and networking. We are seeing a simultaneous strengthening of both international and local ties.

The decade concerned was a period of pronounced progress in Finnish rural policy. National policies gained a new impetus from the broader European perspectives, and European regional development strategies were reflected in Finnish rural development practices. Some of the new means of development fit in excellently with the Finnish scale of operating, for example, the new local action groups and partnerships which represent important innovations in the context of the new rural policy.

The Main Features of Socioeconomic Changes in Rural Areas

The early history of Finland is a history of an agrarian society. Space for settlement was cleared out of the forests and the people made their living from the land and water. The image of a successful citizen was that of a peasant farmer who was independent and self-sufficient. Even after the Second World War there was a firm belief in independent peasant life and the family farm. The widely distributed production of foodstuffs and an even spread of settlement over the country were looked on as safeguards against times of crisis, and this remained the national strategy for survival for a considerable time. It was only in the 1960s that the post-war resettlement schemes finally came to an end. The relocation of farming families from the ceded areas of Karelia was the social policy of the day, but it was also a means by which rural settlement and agricultural production could be committed to the principle of small-scale operation.

Lakes, most of them very small, constitute about 10 per cent of Finland's total land area and forests 70 per cent. The Finns have always had a respect for the forests, and this aspect of life came to benefit from the economies of large-scale operation at an early stage and expanded over the decades into a major, world-ranking industry. The forests continue to provide a significant foundation for economic activity in the rural areas, but the contribution of the related industries to employment in these areas has declined sharply with the mechanisation of timber harvesting. The forest and metal clusters nevertheless formed Finland's most significant branches of international industrial production until the breakthrough of information technology in the late 1980s. Nowadays, 80 per cent of the rural population make their living from sources other than primary production.

The Official Definition of 'Rural'

Finland was an entirely agrarian country until the 1960s, and the 'countryside' was so taken for granted as the background to everyday life that people were scarcely aware of its existence. It really only came to be defined as such with the emergence of urbanisation, when it was recognised as the antithesis of the towns. The towns represented progress and the countryside and its villages the old, underdeveloped order. The abandonment of the countryside had begun, both psychologically and in practical terms.

The rural areas of Finland have always been closely associated with agriculture, and a powerful notion has existed that it is composed of forest and fields and inhabited mainly by farmers and a few forestry workers. Statistically, it has been defined as comprising all the territory lying outside the built-up, densely populated, areas and has been characterised by population loss, with very severe out-migration problems in the 1960s and 1970s, marking a change in the spatial distribution of population that was far more pronounced than anything that took place in nearby Sweden, for example.

It has proved impossible to reshape the Finns' image of the countryside through the medium of research, as little research has been carried out concerning it, nor have any attempts been made to define it as an entity. The rural image that prevailed up until the last decade was the one preserved in the minds of the members of the 'boom generation' who had moved into the towns, while the image harboured by those in power continued to be one of the 'dying villages', the places that they had left behind more than thirty years before.

The roots of rural research and rural policy in Finland go back about 15–20 years, the principal aim of both being to introduce new dimensions into the Finns' understanding of the rural. The two branches of study have developed side by side since that time, the research aspect generating a large amount of new data on the nature of the Finnish countryside. The rural policy branch has worked over the course of the last decade to develop a five-level regional classification of municipalities (Malinen et al., 1994), which was adopted in the summer of 2000 as a basis for the targeting of rural development funds and evaluation of the Objective 1 programme. The five categories recognised in this classification are.

1   towns and cities (main urban areas) (40);
2   other towns and cities (18);
3   economically integrated municipalities (84);
4   intermediate rural municipalities (181);
5   remote rural municipalities (129).

In order to promote interaction between the towns and the countryside, committees for rural and urban policy have proposed a system for categorising local government districts (municipalities) in which 45 are classified as urban, 148 as areas of urban-rural interaction and 310 as rural, the latter including 50 that also belong to the interaction zone.

The most recently published study of rural changes in Finland is *Finnish Rural Areas at the Beginning of EU Membership – Rural Indicators* (Palttila and Niemi, 1999) produced as part of the OECD Rural Development Programme. This identifies rural areas in three ways. Taking administrative boundaries as the basis, sparsely populated areas and small localities of under 500 inhabitants are classified as rural areas and localities of at least 500 inhabitants as densely populated areas (for reference purposes). The data apply to 1990–97.

The authors conclude that about one million Finns live in rural areas, and that patterns of demographic development differ markedly between the rural and densely populated areas. Where deaths have regularly exceeded births in rural areas over the whole period examined, the situation has been the reverse in the densely populated areas. It is also noted that out-migration from the rural areas to other parts of Finland overtook in-migration in the mid-1990s, accelerating the depopulation of the rural areas.

The rural and densely populated areas also differ distinctly in their population structure. The proportion of elderly persons, i.e. aged 65 and over, is higher in the rural areas and the proportion of men is higher in the age groups 25–64 years. On the other hand, the proportion of persons forming part of the labour force within the population of working age is distinctly lower in the rural areas, leading to a higher economic dependency ratio. By contrast, there are no significant differences in unemployment rates between the rural and densely populated areas.

Clear differences between these two types of area become visible when we examine trends in employment opportunities from the end of 1993 to the end of 1996. The number of jobs in the rural areas of most subregions decreased, whereas those in the densely populated areas of most subregions units increased, most job opportunities throughout the 1990s having been created in urban areas. On the other hand, commuting also increased in importance, so that by the end of 1996 up to 45 per cent of employed rural inhabitants were commuting to work in the densely populated areas.

The level and structure of the output of regional economies, and similarly regional economic development, may be described by means of the regional GDP and the value added in industry. The same research referred to above (Palttila and Niemi, 1999) examines each Finnish region in turn and groups its local government districts into rural, semi-urban and urban ones. The

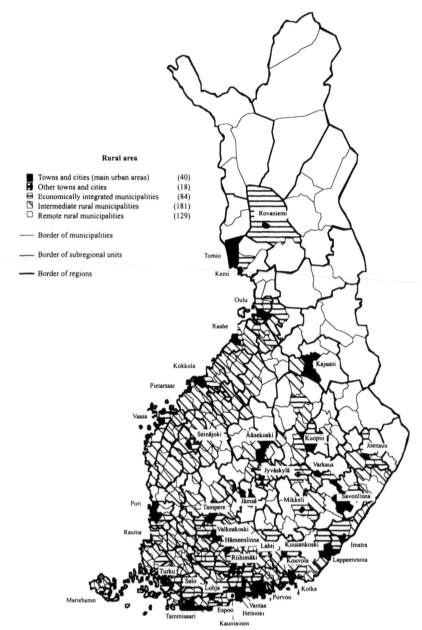

Figure 10.1 National rural classification of Finland at the municipal level

Source: Research and Development Centre of Kajaani; Finnish Regional Research FAR.

number of enterprises relative to population is about the same in the rural and urban districts, but a typical feature of the rural ones is the higher proportion of small enterprises, on account of the fact that two out of every three active farms are located in rural local government districts. The number of active farms in any case decreased evenly between 1990 and 1997 (from 129,000 to 90,000), at the same time as the average area of arable land per farm increased. Thus the area of cultivated land decreased by only 3 per cent from 1990 to 1996, with 66 per cent of this land located in the rural districts.

In 1996 average annual incomes were distinctly lower in the rural areas (€12,353) than in the densely populated areas (€15,966), and the proportion of the population who had completed tertiary education (upper level) was less than half of that in the latter areas (6.5 per cent compared with 14.6 per cent), although the proportion had increased in both.

Although nearly half of the number of hotels and other enterprises offering accommodation recorded in the statistics were located in rural districts, less than a third of the total of overnight stays took place in these relatively small establishments. The relative growth in the number of nights spent in them from 1993 to 1998 had been greater in the urban districts than in the rural ones.

The term 'summer residents' refers to the total number of persons in the household-dwelling units of holiday residence owners. The number of summer residents only includes those whose vacation residence is not located in their home municipality. Three out every of four summer residents, i.e. members of the households of owners of leisure-time dwellings in a local government district where they are not permanently resident (total 536,000 persons, occupying 429,000 summer cottages in all), are located in rural districts. The proportion of summer residents had been growing throughout period examined, and by the end of 1997 they accounted for an average of 32 per cent of the regular population of the rural districts.

The Structure and Nature of Finnish Public Administration

The hierarchical system of public administration in Finland has frequently been described as extremely tightly bound to a set of norms and bureaucratic in a negative sense. It is also sharply divided into sectors that vie with each other to protect their own interests and gain adequate shares of the public funds available.

Certain gradual changes have nevertheless taken place over the last 20 years or so, in that the state administration has been relaxed and more responsibility granted to the regions and the local authorities. The last administrative reform (1997) cut the number of provinces to six and greatly reduced the role of their administrative bodies in matters of regional development. Thus, as far as rural development is concerned, it is more relevant to look at the administrative changes that have taken place in the regions, and particularly at the local level. The Provincial Government of the Åland Islands is responsible for regional development matters as well.

Regional and Local Government Structures

Work on restructuring the system of regional administration continued throughout the 1990s, involving two main approaches that deviated slightly from each other. In the first place, there is a system of 20 Regional Councils with statutory responsibility for regional development and operating under the political control of the local councils in their areas. The political interests of these regional councils come close to those of the Centre Party, which is committed to supporting their administrative role.

Each region has a partnership group that deals with all development activities in the different spheres of government that are co-funded by the EU, and with the individual measures involved in these. The partnership group includes representatives of the region, its local authorities, the state authorities, business and industry, the labour market organisations and other interested parties, and its meetings can be attended by representatives of the Ministry of the Interior and the European Commission.

Meanwhile, the Social Democrats, who wield the greatest political power in the country, have come out firmly in support of a model based on Regional Employment and Economic Centres (*Työvoima- ja elinkeinokeskukset*) which are subordinate to the relevant central government agencies. There are 15 of these in all, and they function as advisory organisations without any connections with representative bodies but under the immediate vertical control of the ministries responsible for the sectors in which they operate, i.e. rural affairs, labour and private enterprise. These units are responsible for administering a very much larger proportion of the public funding for the development of business, employment and agriculture than are the regional councils.

The Significant Role of the Local Authorities

The local authorities in Finland have a significant role in rural development. At present there are 452 local government districts (municipalities) in Finland, including towns, cities and rural districts. These differ greatly, however, in terms of both area and population. Most of them are small, beginning from Velkua, with 235 inhabitants, while the largest in terms of population is the capital, Helsinki, and the largest in area is Inari in the province of Lapland, which nevertheless has a population of 7,450.

The local authorities are expected to promote the well-being of their inhabitants and foster development in their areas. They are able to act independently in many matters, choosing the most convenient ways of discharging their responsibilities, i.e. their contributions, both statutory and optional, to the public services provided for their inhabitants. The most significant functions of the local authorities are in the fields of education and culture, health care and nursing, social welfare and community planning, and they exercise a substantial influence on the lives of their inhabitants, since this means that they are currently taking care of nearly two thirds of all public services.

The system of state subsidies to the local authorities was reformed in 1993 to the extent that the money was no longer earmarked for particular purposes, leaving the local councils to decide how to allocate finance between the social services, health care, education and culture on the one hand and between their remaining responsibilities on the other, taking the local needs for particular services into consideration.

The local authorities are at present searching for new development strategies and thinking more seriously than ever about the benefits of cooperation and networking. One tendency has been to transfer development policy away from local government and to allow new actors to emerge at the local level. The status of local development policy in the rural districts of Finland and procedures adopted by it changed more markedly than ever before during the 1990s, as a result of unprecedented evolutionary forces. One could say that local development policy has never faced such changes since the time when it was conceived (Kahila, 1997, p. 114). The main sources of change were: the new Local Government Act of 1995; changes in local government finances; implementation of local development policy at the sub-regional level; economic networking; the new Regional Policy Act of 1994; membership of the European Union; formation of local action groups; and the development of a national rural policy.

Before Finland joined the EU the development work that took place in its rural areas was mostly governmental, and its implementation largely followed bureaucratic models and accepted administrative procedures. Projects were organised as a natural part of the local administration, with no private sector involvement. The recent tendency in local rural development policies has been a shift of emphasis to the sub-regional level. Rural districts readily adapted to membership of the EU, and the beginning of the work of the local action groups marked a new era in the implementation of local development policy, a departure from the traditional narrower and more sectoral approach (Kahila, 1999).

The numbers of projects of different kinds have increased, but an important difference lies in the fact that the local councils do not produce them by themselves, but instead they are designed and implemented by external actors. According to Kahila (1999), the councils are still adhering to traditional, sectoral forms of administration and have not really taken note of their new operating environment. Local authority plans make mention of endogenous development and local development initiatives, but the implementation is missing because this new way of thinking and acting is still in its infancy.

Local action groups represent a new dimension in development policy. For the first time these are in possession of independent funding at the local government level. The tradition in implementing local development policy in the Finnish rural districts has previously been one of bureaucratic programme planning, and the partnership approach has been less prominent, but the new requirements placed upon local authorities have changed this approach entirely (Kahila, 1999).

Levels of Rural Policy Implementation

Rural policy applies to various levels, concerns both private and public organisations and may be based on either formal or informal structures. Rural development and rural policy become sufficiently precise if their implementers are competent developer institutions capable of promoting independent activity. In principle, rural policy is connected with the three divisions of administration: central government, regional administration and local government.

National regional policy is guided by the Regional Development Act (1135/ 1993), which came into effect at the beginning of 1994, supplemented by the Regional Development Statute (1315/1993). The aim of the act is to promote the independent development of the regions and a good regional balance by means of programme-based regional policy and graduated regional support for business and industry.

The principles of regional policy and its general lines for development, together with the action that falls within the purview of the government, are defined in the Objective programmes. Regional development funding takes the form of specified budgetary allocations to the various sectors of government which can be used to promote regional development targets and the programmes drawn up to achieve these.

One problem from a regional and rural development perspective is that the Finnish administration system is so strictly divided into sectors, whilst at the same time there is a lack of skill in local, regional and horizontal activities and a lack of the structures needed there. The countrywide review of Finnish Rural Policy carried out by the OECD in 1994 regarded the incoherence of administration at the regional level as a weak point in the Finnish system, as it means that the idea of an integrated rural policy has not yet reached the regional level. Instead, the work carried out under the regional programmes has been able to strengthen the subregions.

The original content of the concept of regionalisation is related to the expansion of urbanisation during the 1970s into the rural districts surrounding the population centres while most places of work were still located in the centres. As the numbers of tax-payers in the surrounding rural areas increased, these areas became more powerful, a pattern that was repeated throughout the country. 'Instead of a dividing line between urban and rural areas, the basic division in the community system became that between the urban subregions and the real rural areas outside them' (Vartiainen, 1997, p. 91).

The regionalisation of the local economies has proceeded gradually in parallel with the development of the community structure, created by new collectivities called local partnerships. Local action groups are needed in both the rural and urban subregions, with the sub-region or some similar area as their recruiting ground. Those working in the groups will then be able to solve the problems of their subregions or promote development projects in them.

Communality, which is a crucial part of an integrated rural policy, is a very complicated concept when viewed from the local government perspective, but it does link the local level to the regional one, where there are common threats as well as opportunities and targets, and it allows council representatives to work in local action groups. Communality in regional and rural policy offers the councils more opportunity to exercise influence and to create a forum for direct action, but partnership is required to ensure their autonomy. In practice, changes in operating conditions could mean a continuation of the regionalisation of local authority economic policy.

Village Action

The village is no longer an administrative unit in Finland, as a network of local government districts was set up at the end of the nineteenth century to take care of administration at the local level. The villages have nevertheless always been present throughout the history of the Finnish countryside as significant social communities and regulators of the immediate living environment. The current village action approach is a mode of local participation, of voluntary collaboration between inhabitants for the improvement of their own living conditions, that developed during the 1970s as a means of counteracting the social changes of that time, which were tending to undermine the vitality of rural areas. The model spread across the whole of Finland within a space of 25 years, and has become a part of the development of civil society in the countryside and of the evolution of the tradition of voluntary work (Hyyryläinen, 1999).

The idea behind village action is an internalised concept of the human being as a creative actor, a subject. This emphasises individual people's active relationships to the environment in which they live. The theoretical narrative for this regards human beings as capable of consciously controlling their own conditions and systematically developing their patterns of action. The principal arena for village action has traditionally been the places where people live out their daily lives and the immediate surroundings of these, in other words their own villages and their environs.

Village action has traditionally been closely connected with people's experiences of life, and its general purpose has been to develop the village, increase its vitality and support people's belief in life. Vitality is usually interpreted in a highly functional sense, as a spirit of doing things together. The animation of a village has been understood as affecting its inhabitants' quality of life and enjoyment.

As with many other forms of voluntary activity, the principle on which village action relies is personal unpaid work. This is a very specific concept in Finnish – *talkoot* – which is a form of voluntary cooperation governed by generally agreed norms and based on local networks of mutual trust. Thus village action in its present form can be looked on as a stage in the historical evolution of this voluntary work by which the original reciprocal cooperation in agricultural production has developed into a modern form of voluntary activity and eventually into a means of achieving local development through communal self-help (Hyyryläinen, 1994, p. 134). It is crucial to note that the value of *talkoo*-work was acknowledged officially at the beginning of the Finnish LEADER II programme.

The most common form of organisation adopted for this purpose is the village committee (*kylätoimikunta*), a voluntary group constituted in a manner agreed on jointly by the inhabitants themselves. Thus in its most typical form it is not a judicial body but an entirely unofficial actor, or 'creative subject'. Other possible forms that a village committee can take are a developmental association or cooperative or some other registered society that has adopted the same principles.

Village activities have come to a stage where they could take the main responsibility for development work in the villages. Naturally, the task demands cooperation with numerous organisations, but the village committees and those comparable with them are best acquainted with the present situation of the rural areas and the programme policy. They are the planners of the development programmes for the villages and take part of the responsibility for the implementation of village projects. At a certain stage, the achievements of these activities will go deeper into the population base of the village and release new power for non-governmental organisations.

The POMO Programme: A Finnish Supplement to LEADER

The LEADER Community Initiative Programme in Finland was under the jurisdiction of the Ministry of Agriculture and Forestry. Finland had two LEADER programme documents in 1996–99, one for Objective 5b areas and the other for Objective 6 areas, covering 43 per cent of the inhabitants of the former areas and 42 per cent of those of the latter.

One important target for Finnish rural policy has been to expand the LEADER type of policy to the whole country. An actual proposal to this effect was made in autumn 1995, when the LEADER programme was not yet ready but its notable activating effects had already been appreciated. There had been much talk of subregional development, and it was estimated that such an activity would cost about FIM 45 million annually. An attempt was made to obtain the money from the state budget in 1997, but after several rounds of discussions, only FIM 5 million was granted, even though both the government and parliament had approved the idea.

In addition to the budget, the idea was incorporated into the government's decision in principle on regional development announced in winter 1996 and in the government statement on the regional development programme submitted to parliament in winter 1997. When it became obvious that there was a chance of obtaining more money from the agricultural development fund, the Rural Policy Committee prepared its sub-region programme document in winter 1997. This programme was named POMO, an acronym derived from its name in Finnish *Paikallisen omaehtoisuuden maaseutuohjelma*, which may be roughly translated as the 'Rural Programme for Local Initiative'. The subregional aspect includes the planning and implementation of strategies initiated at the grassroots level and projects similar to those contained in the LEADER programme. The idea was that POMO should strengthen the local initiative structure and subregional development work.

POMO activities started in March 1997 with the distribution of information and programme documents. The subregions attached to the programme were requested to send their three-year plans for the period 1997–99 to the sub-regional development theme group at the Ministry of Agriculture and Forestry by the end of July 1997, after which this body would prepare the Ministry's decision on the formation of POMO action groups. A sum of FIM 28 million and a set of rules for running the POMO project were received in the course of the summer. The scarcity of grants meant that it was not possible to expand the policy represented by the LEADER method to the whole country, but the 26 action groups which were selected in September 1997 nevertheless covered one-third of Finland's rural areas.

According to the original proposal, the POMO fund established by the Ministry of Agriculture and Forestry represents a half of the total cost of the programme, the other half remaining to be collected from other sources, so that at least 25 per cent of the money will either be of private origin or will ensure the commitment of the local councils.

Only the LEADER areas and the large population centres were left out of the POMO programme, whereas the peripheries of urban areas, for example, were included, as in the case of LEADER. As for measures which promote interaction between rural and urban areas, it is still possible to expand the areas affected by these in the direction of the towns. The programme document gave detailed information on the purpose of adopting the POMO programme, regarding it as a method for strengthening local independence in regional development during the coming EU programming periods. The aim was that POMO should be a better and easier programme than LEADER while still guaranteeing the autonomy of its action groups.

For a local group, the POMO programme is the most crucial and important operator in local development. It is responsible for implementation of the work and covers the expenses. It has also had to organise itself as a legal association. The most essential issues, however, are the openness of the action groups and their work and the confidence placed in these groups, which are responsible for the planning, execution and promotion of the programme and the dissemination of information on its activities in the areas where these take place. Implementers and activities are selected by the POMO groups themselves, which are responsible for the administration of the activities and for maintaining contacts with the officials to the extent necessary for the projects. All measures should be subject to the appropriate laws (POMO Programme Document, 1997).

The Rural Policy Committee selected the local development theme group to prepare the programme and guide the process of its work, and the theme group then prepared the selection of action groups, determined the financial framework, prepared the Ministry's agreement along with the above-mentioned groups, spread information on the development of the POMO programme, provided the necessary training and developed the system of payments. The purpose was to create an easier way of proceeding, in which the Ministry of Agriculture and Forestry determined the action groups' annual POMO budgets on the basis of their contributions to the programme and their financial plans. The POMO programme for each group has the status of a project in this process, and each act of allocating funds to a specific target is a sub-measure. The Ministry's right of inspection nevertheless extends as far as controlling these sub-measures.

In order to intensify the communication of information and activate local development work under the programme, the theme group nominated two activators from within its circle. The selection of activators proved a success, and information was communicated successfully in real time between the theme group and the action groups that were in the process of being established. Significantly, this took place in both directions. In the initial phase of the programmes, there were intensive programme policy demands for the dismantling of the central administration. In this way it was possible to minimise many misunderstandings which could have had long-term effects and slowed down the activities.

It was obvious from the activators' report (Ehrstén and Niemi, 1997) that the arrangement of funding from many other sources, for example, and traps related to the role of the local councils had successfully been avoided through constant interaction between the central administration and the local level. By virtue of their activity, the animateurs had partly been able to reduce the erroneous bias in the targeting of the POMO funding, after which it became possible to proceed to the main point of building up the necessary programme and preparing measures relevant to the activities.

Beginning from 1998, the Finnish Village Action Association began to function as a network unit for the POMO groups. For some time there was a vast village network project in operation in the same office, the goals of which were similar to those generally required of network units, e.g. production of their own newsletter. The purpose was to unite activities of the same kind in the subsequent programming periods.

Although the POMO programme remains a two-year exercise so far, this does mean that the action groups can get on with their activities, and that the experience gained and lessons learnt will have considerable effects on the application of LEADER+ during the next period. In addition to the long-term objectives, and in spite of its short period of operation, the POMO programme has produced successful results. Moreover, it has served as a catapult for the preparation of projects for other programmes.

The intermediate reports on the Finnish LEADER II process (1997) and the analysis performed by Kirsi Viljanen (1997), which in both cases contained many interviews with implementers of the community initiative, emphasise the bottom-up operational pattern which is followed in regional development planning schemes and the principles of locality and partnership. Similarly it is clear that the equality aspect, programme additionality and the principle of sustainable development will also be taken into consideration.

Many administrative, nonadministrative and political problems were encountered within the activities of the LEADER and POMO programmes when it came to finding means of implementation corresponding to objectives and ideas that would be in accordance with EU regulations and which at the same time would pay respectful attention to Finnish administrative practicalities. It seems to be impossible to find a solution to this dilemma, but as far as the two programmes are concerned, although there are certain issues that can be improved, it has still been possible somehow to get on with the activity itself. An outsider might presume that the main tension would be created between the member state and the EU, but this has not been the case at all.

As far as Finland is concerned, the tension is related to the internal division of labour between the central and regional administrations and the process of making local decisions. The Finnish administrative pattern traditionally follows sectoral and top-down principles, whereas the basic objective of a community initiative is to activate people, entrepreneurs and producers to join in local and regional development work. An administration system which is an instructive, decentralised, authoritatively persistent distributor of money according to a certain quota and committed to supervising the outcome has no resources for action as such.

It was quite surprising for the Finnish administration system that the principles of the EU Community Regional Policy and those of subsidiarity and partnership extend as far as the practical administration and therefore

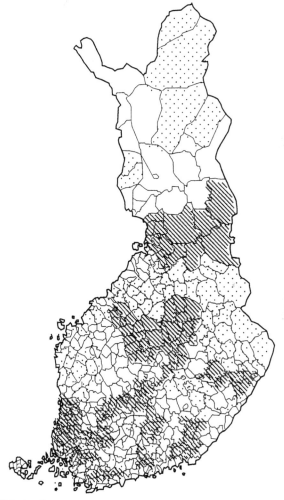

☐ LEADER II Action Groups

▨ POMO Action Groups (Programme of Rural Development based on Local Initiative)

Comparison between local rural action groups in Finland 1997–99

Programme	Inhabitants/area (average)	Members of LAGs	Projects until 7/99	Total funding (in million EUROs)	Duration of Programme
LEADER II	37,000	390	1,900	76–84	9/96–12/99
POMO	35,000	470	1,400	28	12/97–12/99

Figure 10.2 LEADER II and POMO Action Groups in Finland 1999

Source: *Finnish Journal of New Rural Policy*, 1999, p. 97.

have to be taken into consideration. In this way the traditional line of administration had to be bent towards a new position. It is predictable that practicalities which are driven by community initiatives will be applied more broadly and the methods will also be developed further.

According to the latest research the implementation of an integrated rural policy has had significant impacts at many levels. Pressures for change, all related in one way or another to social interaction and collaboration, have arisen both in the forms of cooperation pursued by ministries and regional authorities and in the organisation of joint human activities at the local level. The implementation of an integrated rural policy in Finland has met with many structural problems. A certain amount of friction has been experienced in incorporating the key principle of action within the EU, that of partnership, with the forms of organisation and action typical of the Nordic welfare state, and similarly there are many private, public and community actors operating in the regions, and it will be important with a view to the success of rural policy that these should not simply concentrate on developing their own capacities but should foster new efforts to develop intersectoral capabilities (Hyyryläinen and Rannikko, 2000).

Discourses of Rurality and Rural Policy

The most extensive discussions of rural policy took place 10–15 years ago, and by the late 1980s, the numerous detrimental factors entailed in the methods used for rural development had been noted (Uusitalo, 1999). A concise rural policy was needed as a precision weapon for counterbalancing purposes. This was set out in the Active Countryside programme (1996, pp. 22–3) in the following manner:

1   special questions concerning rural areas are obscured by the interests of a sectional administration, which means that contact with the whole entity has vanished, leading to a reduction in rural functions and urbanisation, i.e. a movement towards the conditions prevailing in the centres;
2   regional policy has become more detailed and regional discussions have collapsed as different measures have proved to have contradictory impacts;
3   the value attached to rural areas in the context of development activity has increased. The countryside is now regarded as a thing of value in itself and a resource for development which has a lot to offer of its own in terms of social policy.

*The Active Countryside Campaign and Rural Development Project
1988–91*

In 1988, Finland along with 24 other countries took part in a rural campaign initiated by the Council of Europe. The message of the campaign was meant especially for young people, women and urban dwellers, and immense efforts were expended on these target groups In Finland a total of 29 national organisations, the Association of the Regions and more than 200 local authorities participated. The 10 principles of rural development set out by the Finnish Rural Campaign Committee have been fulfilled to a large extent (The Active Countryside Campaign, 1989, pp. 22–3):

1 attention should be paid to the countryside's non-material resources;
2 constructive activity should be based on the prevailing cultural environment;
3 an active countryside consists of active communities;
4 special attention should be paid to the status of women and young people in rural areas;
5 work should proceed from sectors to the development of regional entities;
6 regional differences between rural areas should be taken into consideration in planning and development, with particular attention to remote areas;
7 the accent should move from the public authorities to local initiatives;
8 the barriers between the public and private sectors should be abolished;
9 economic development should proceed from the natural environment in the countryside, i.e. the cultural inheritance and rural raw materials;
10 the necessity for a shift from agricultural policy to rural policy must be appreciated.

The Campaign was a success, and partly inspired by this, a Rural Development Project was implemented in 1988–91 through cooperation between the Ministries of the Interior and Agriculture and Forestry. This project was functionally a more systematic and sustainable way of carrying out rural development work. The time was right for undertaking new forms of activity. The advancement of the campaign, which was a tangled matter to some extent, eventually served as an incitement to pursue the issue.

The rural project became bogged down to some extent. There was no question of producing a report, as there was no money available for this in the first place, nor was there sufficient need for giving statements. The question of what was to be done next was an entirely new one for the group nominated by the state administration. The solution was to prepare an overall programme for rural policy. This required taking advantage of the various small working groups and bodies of experts that had been formed, the inputs from which resulted in the production of publications which were considered to be of considerable interest.

The rural programme issued 79 proposals, which were regarded at that time as fairly concrete. Evaluation of the project later on proved that the projects were not quite as expected, but it did point to more straightforward, responsible measures than people had been used to before. Implementation of the proposals did not become a contribution of the rural programme as such, however, and only 14 proposals were totally implemented and 26 partly. In particular, the proposals concerning taxation failed, on account of the old-fashioned attitude adopted by the rural project towards the taxation system, which failed to take into consideration the process of tax reform that had already been initiated.

New Forums for Administrative Partnerships

After the rural development project the Rural Policy Advisory Committee (*Maaseutupolitiikan neuvottelukunta*) was appointed (until 1995). The composition of this committee was determined basically according to the significance of rural policy for different sectors of the administration and other organisations. In order to promote the idea, a member involved in a coordination function was required to be absolutely committed to his or her task and role. This still created problems, however, in that the traditional custom is merely to serve as a representative of one's own organisation to discuss matters of common interest, i.e. members are passive observers rather than participating actively. The time limit for the advisory committee was decided by the government on the basis of the estimated length of time needed.

During the life of the Rural Policy Advisory Committee, preparatory work in the field of rural policy was organised into theme groups. Project activity increased considerably, and research funding was obtained in 1992. On the whole, the advisory committee was working under difficult circumstances, in that preparations for membership of the European Union caused uncertainty as far as its status and objectives were concerned. In addition, it was badly criticised as worthless due to its attachments to political parties. It was said that the EU will bring with it solutions and forms for rural development work, a Common Agricultural Policy and structural funds that will make their impact through the means of regional development.

The Rural Policy Advisory Committee prepared the promised Government Rural Policy Declaration and submitted it in October 1993, but before that certain problems of a political nature arose. In actual fact, rural policy was emerging as a political matter. The political dilemmas that came to the fore during 1993 and 1994 in the development of rural policy may be attributed to the effort made to strengthen this policy in a manner amenable to the political administration. Until then it had been a harmless project or campaign and on the political level it had remained part of regional or agricultural policy. The newly drafted rural policy and the existence of the advisory committee placed it in various new fields of administration, a situation which required cooperation between them.

This was backed up politically by the government's declaration. The hidden dispute concerning responsibility for rural development had to be solved, although from the point of view of the traditional sectors of administration it was regarded as a 'no-man's land'. As is commonly known, genuine collaboration as a working method is rare in a ministry, and is partly impossible on account of the administrative regulations, and therefore the delay was understandable even though not acceptable.

Although the statements of the declaration were based on a rural programme compiled a couple of years earlier, they did not actually contain any concrete proposals. They did, however, specify a set of rural policy activities, these being modernisation of sources of livelihood in rural areas; strengthening of the operation of the already established service network; developing of the quality of living conditions and the community structure; and sustainable utilisation of renewable natural resources.

The difficult working atmosphere during the period of the Rural Policy Advisory Committee reflected upon the process of nominating a successor to it. Its secretariat had already prepared a proposal in the summer of 1994 for the formation of a Rural Policy Committee (*Maaseutupolitiikan yhteistyöryhmä*, YTR) to succeed it in its work, in the hope that the name would go hand in hand with the real nature of the work that it was expected to focus on. The Advisory Committee had come to be regarded within the state administration merely as a body that issued statements, which was utterly insufficient for taking care of rural matters, even at a time when the sectoral administrations were still handling their own aspects of rural activities. In order to seek a political way of strengthening rural policy, the Rural Policy Committee was to be set up by the Council of State, even though its members were in practice nominated by a single ministry.

The Rural Policy Committee (YTR) is a working group formed by civil servants. Compared with the composition of the previous working group, it is clearly intended as a means of collaboration between different sectors of the administration, a forum in which nine experts from different ministries and six expert organisations or interest groups endeavour to find joint solutions to problems that are common to them all, even though they may be either marginal or completely insurmountable for a single organisation.

It had become obvious that there is need for constructing regional project plans partially in cooperation with the European Union's fund and for implementing projects which have been enlarged from an initial plan. Towards the end of 1995, implementation of the Active Countryside project was undertaken by the Rural Policy Committee. Meanwhile, activity under the national Rural Policy Research and Development Project had made further progress. Finally, conditions seemed to be right for changing the policy of devising vast, concerted projects. In the summer of 1995, the government of Prime Minister Paavo Lipponen appointed a Working Group for Rural and Urban Policy programmes, although there also existed an independent Working Group for a Rural Policy Programme.

The Second National Rural Policy Programme and EU Membership

It has been observed that rural policy has gradually become established in the course of the Active Countryside programme *(Toimiva maaseutu-ohjelma)*, which describes the functions of rural policy more accurately. What is new is that the programme affects the starting points for rural policy even when these are totally at variance with certain provisions made in the field of traditional politics.

Such contradictions exist in almost every sector of the administration, but most of the cases become apparent in the environmental administration, or at least had done so by the end of 1990. Typical of this second Overall Rural Policy Programme was the greater concreteness of the proposals. This together with the strengthened preparatory group made it possible to implement its fairly good proposals. Out of the 96 proposals, 80 had been totally or partially implemented by June 1999, while the remaining 13 were in the process of implementation. It is anticipated that most of these will also be implemented in the course of time. Additional money was obtained to carry out the implementation, but not the FIM 0.5 billion which had initially been applied for (Uusitalo, 1998, pp. 63–6).

The essential changes made in the content of the programme compared with the previous one were the strong emphasis on local and regional aspects, the concrete proposals made in connection with this and the view expressed on the importance of networking. Cooperation was enlarged to farms and other rural enterprises as well as administrative organisations and expert systems.

Joining the EU caused uncertainty for a couple of years on how to organise rural policy. The functions of the Rural Policy Committee and the Union's programme involving various groups of officials were regarded as overlapping, although others considered them complementary, as the two instances had several functions in common. The latter view was proved right in 1997, when the Committee discussed the possibility of accelerating implementation of the programmes and the theme groups became aware of their position as servants of several regional projects and promoters of common tasks. The EU programme brought in a lot of additional money for rural policy, and in the meantime the tempo was speeded up significantly both regionally and in a number of specific occupational fields. Agenda 2000 and the continuous changes in means of livelihood and administration structures have nevertheless led to a permanently delicate and tense operational environment.

It was understood in the autumn of 1995 that rural policy is an essential part of the financing of projects and theme groups. For this reason it was possible to increase and enlarge activities such as small and medium-sized wood processing enterprises, rural tourism, small and medium-sized food processing companies, telework and village activities. An estimated 4,000 jobs were created as a result of the above entrepreneurial efforts and many existing jobs were made more secure. The number of jobs was expected to rise to at least 8,000 over the following three years. This included to some extent the achievements of projects partially assisted financially by the EU. The preparation of a rural policy which lies outside the partially EU-financed programmes has cleared the way for such projects to start. It is also reasonable to include in the list the real development work that has taken place with regard to natural products, which is powerfully assisted by rural policy project funding.

The main objects of rural policy are the rural people, regardless of aspects such as profession, gender or age. It can be said that the restricted rural policy has encouraged the people who live and work in rural areas, has given impetus to a growing branch of industry and has paid close attention to the cumulative worsening situation with regard to services in rural areas. The system is not complete and it is extremely sensitive, but since a number of active elements have been created, it is expansive by nature. Rural policy is neither a project nor a pilot. In fact it is a permanent operation, the objectives and measures of which are not defined according to the boundaries between sectors of either the administration or the economy.

The Third National Rural Policy Programme 2001–03

The Third National Rural Policy Programme is now in preparation, in a more favourable political climate than its predecessors. Rural problems are now able to elicit a response in both administrative and political circles, and it is particularly significant that the rural policy perspective has broadened into a social policy one at this preparatory stage. The programme is aimed at achieving a new balance between sectoral policies and programme-based rural development.

The Third Rural Policy Programme is in line with the theoretical precepts of European rural policy, with central emphasis placed on the integrated rural policy paradigm and its practical implementation. Rural development cannot be brought about without integration of the objectives and expertise of various branches of the administration, achievable in practice through horizontal cooperation within central government together with regional and local partnership. One important form of activity in an integrated policy is programme work, and one major instrument in this is innovative project activity.

Rural Policy as the Herald of a New Political Culture

The beginning of the new millennium can be looked on as a significant time of exploration as far as the future of the Finnish countryside and rural policy is concerned. In the first place it must be admitted that the vitality of the rural districts has been drained almost to the point of exhaustion, so that although sustainable development of the regional structure is still possible, it is clear that its independent basis for development cannot be allowed to dwindle any further without the danger of expensive reliance on external resources for the maintenance of viability. Secondly, this is a period of exploration in the sense that the beginning of the new decade seems to be bringing with it alterations in values and attitudes which may be of benefit to the vitality of the countryside in Europe as a whole. The new millennium will evidently herald new lifestyles and changes in attitudes, so that this is the time for making the political choices and investments required in order to achieve a sustainable rural policy.

Although trends in rural policy in Finland have been largely a matter of administrative consolidation, the development of thought patterns and publication, some practical reforms have been carried through that had previously proved impossible in the field of employment policy, for instance, in spite of the many experiments undertaken (Katajamäki, 1998; Luostarinen and Hyyryläinen, 2000). As a concept and principle, partnership has been written into the sectoral development plans of many branches of administration, and it has gained concrete representation within rural policy to a more systematic degree than in any other sphere. One practical demonstration of this is the fact that the new LEADER+ programme has attracted applications from a total of 57 local action groups. This means that such groups cover virtually the whole of the Finnish countryside and that this mode of action has gained widespread acceptance.

The practical tools for achieving development on a local scale have not yet developed to become sufficiently diversified or of sufficiently high quality, nor have the meagre resources available for the implementation of rural policy been able to create opportunities for developing the work of the local action groups, e.g. in the form of methods for carrying out and administering programme and project work. Considerably more effort will have to be devoted to such aspects in the future.

The Finnish countryside cannot be rescued with programmes and projects alone, however, but rather policies and strategic decisions will be required at the national and international level. The role of the EU's programme policy will need to be filled out and renewed, so that it can provide opportunities for developing rural areas of different kinds, especially those in more remote locations. We should be prepared to learn from earlier experiences when exploiting resources, although bearing in mind at the same time that the flow of cash into the structural development of the existing member states is not likely to continue at the current level for very long.

It is essential to note the connection that exists between rural development and the particular features of the country concerned and its natural competitive potential. Finland is a very sparsely populated country with small towns and vast rural expanses, so that these rural areas and the scale on which they occur may be regarded as part of the country's character. On the other hand, they are inevitably bound up with the general structural trends taking place in Europe, trends that will unite the people of Europe in quite unexpected ways. Apparently small beginnings and associations could well develop in the future into important points of similarity or attractive narratives and thereby give rise to political movements. We are living in a time of increased awareness of the similarities of the problems affecting the Finns, the French, the Irish and the other nations of Europe.

References

Active Countryside Campaign (1989), *Maaseutukampanjan Loppuraportti*, Valtion Painatuskeskus, Helsinki.

Active Countryside Programme (1996), *Toimiva Maaseutu-ohjelma. Maa- ja Metsätalousministeriön Asettaman Työryhmän Ehdotukset ja Perustelut*, Maaseutupolitiikan yhteistyöryhmän julkaisu 1/1996, Helsinki.

Ehrstén, L. and Niemi, R.M. (1997), *POMO-ohjelman Aktivaattoreiden Raportti*, Maaseutupolitiikan yhteistyörhmän seudullisen kehittämisen teemaryhmä, Helsinki.

Hyyryläinen, T. (1994), *Toiminnan Aika. Tutkimus Suomalaisesta Kylätoiminnasta*, Line Sixtyfour Publications, Tampere.

Hyyryläinen, T. (1999), 'Changing Structures of Finnish Village Action', *New Rural Policy*, vol. 99, pp. 98–105.

Hyyryläinen, T. and Rannikko, P. (eds) (2000), *Eurooppalaistuva Maaseutupolitiikka. Paikalliset Toimintaryhmät Maaseudun Kehittäjinä*, Vastapaino, Tampere.

Intermediate reports on the Finnish LEADER II process (1997), *Suomen 5b ja 6 – LEADER-ohjelmien Väliarvioraportit*, Suomen Aluetutkimus, FAR, Sonkajärvi.

Kahila, P. (1997), *Kolme Maata ja Kolme Elinkeinopolitiikkaa: Suomi, Saksa ja Irlanti*, Turun yliopiston julkaisuja C 138, Turku.

Kahila, P. (1999), 'Local Development Policy in Rural Municipalities', *New Rural Policy*, vol. 99, pp. 75–80.

Karppi, K. (2000), *Articulated Spaces*, University of Tampere, Department of Regional Studies and Environmental Policy, Tampere.

Katajamäki, H. (1998), *Beginning of Local Partnership in Finland. Evaluation, Interpretation and Impressions*, Research Institute at the University of Vaasa, Publication No. 76, Vaasa.

Luostarinen, S. and Hyyryläinen, T. (eds) (2000), *Uudet Kumppanuudet*, University of Helsinki, Mikkeli Institute for Rural Research and Training, Publication No. 70, Mikkeli.

Malinen, P., Keränen, R. and Keränen, H. (1994), *Rural Area Typology in Finland: A Tool for Rural Policy*, University of Oulu, Research Institute of Northern Finland, Research Report 123, Oulu.

New Rural Policy (1999) Finnish Journal of Rural Research and Policy. English Supplement. (www.mtkk.helsinki.fi/mua.html).

Palttila, Y. and Niemi, E. (1999), *Suomen Maaseutu EU-kauden Alussa – Maaseutuindikaattorit*, Katsauksia 1999/2, Tilastokeskus, Helsinki.

POMO programme document (1997), *Seudullinen Kehittäminen/ Seutuohjelma- POMO. Ohjelma-asiakirja 12.3.1997*, Maa- ja metsätalousministeriö & Maaseutupolitiikan yhteistyöryhmä, Helsinki.

Pyy, I. and Lehtola, I. (1996), 'Nordic Welfare State and Rural Policy', *New Rural Policy*, vol. 96, pp. 17–34.

Uusitalo, E. (1998), *Maratonia Suossa. Toimiva Maaseutu – Ohjelman Laatiminen ja Toimeenpano*, Maaseutupolitiikan yhteistyöryhmän julkaisu 3, Helsinki.

Uusitalo, E. (1999), *The Basics of Finnish Rural Policy. Vitality for Rural Areas. Why, for Whom and How?*, Maaseutupolitiikan yhteistyöryhmän julkaisu 6/1999, Helsinki.

Vartiainen, P. (1997), 'Maaseutuyhteiskunnasta Monikasvoiseen Kaupungistumiseen', *Suomi 80. Kuokasta kännykkään*, Suomen Kuntaliitto, Helsinki.

Viljanen, K. (1997), *EU: LEADER II-yhteisöaloitteen Toimeenpano Suomessa*, Maaseutupolitiikan yhteistyöryhmän julkaisu 5/1997, Helsinki.

Chapter Eleven

Rural Development and Policies: The Case of Post-War Norway

Nina Gunnerud Berg and Hans Kjetil Lysgård

Introduction

In Norway, research on regional development is regarded as virtually synonymous with research on rural development. For example, the Norwegian Research Council's recent research programme on regional development stated explicitly that the main focus of the programme would be rural Norway, defined as all parts of Norway except the conurbations around Oslo, Bergen, Trondheim and Stavanger (NFR, 1999). Similarly, regional political discourses have a rural profile, and have focused on rural problems for most of the post-war period.

Our main focus in this chapter is on rural development trends and rural policies. Because it is difficult to analyse these without reference to urban development and policies, the main themes such as changes in settlement patterns, economic restructuring and policies, are analysed from an urban-rural perspective. Although many issues have been part of both academic and political discourses on rural change since the second world-war, migration from rural to urban areas has been considered the key problem, leaving some rural areas depopulated to varying degrees. Accordingly, rural–urban migration provides this chapter's point of departure as well as concluding focus. This issue is addressed using both quantitative and qualitative approaches.

There are five main sections to this chapter. In the first section we describe changes in settlement patterns and dominant migration flows in post-war Norway, while our main focus in the second section is regional economic restructuring in the same period. The third section provides a short description of the administrative structure of contemporary Norway. In the fourth section we give a brief summary of regional policies since the Second World War. In the final section we examine some dominant social representations of rurality and urbanity, and suggest that they tie into processes of regional development – not least migration flows – in ways far underestimated in Norwegian regional development research and policies.

Settlement Patterns and Migration Flows

The mainland of Norway covers a relatively large geographic area (323,877 km^2) and has a relatively small population (4.4 million). The number of inhabitants per km^2 is only 14.5, which is very low compared to other countries in Europe like Great Britain (241 per km^2), Germany (230 per km^2) or the Netherlands (374 per km^2) (SSB, 1999). Norway is far more scarcely populated than countries in central and southern parts of Europe. Norway is also scarcely populated compared to other Nordic countries. Only Iceland with 3 inhabitants per km^2 has a lower population density than Norway. Sweden has 20 per km^2 and Finland 15 per km^2, while Denmark is more in line with the continental EU countries with 122 per km^2 (SSB, 1999). The number of inhabitants per km^2 says nothing about the pattern of agglomeration within countries. Norway, for example, is characterised by low population density in the most northerly part and high population density in the far south. The density is especially high in and around Oslo, the capital.

There is no official definition of urban and rural areas in Norway, only a definition of densely and scarcely populated areas. A densely populated area is an agglomeration having at least 200 residents and where the distance between houses does not exceed 50 metres.[1] Areas outside this definition are defined as scarcely populated. The definition of a densely populated area differs in one respect from corresponding definitions in Sweden and Denmark. The lower limit of 200 residents is the same, but in these countries the distance between the houses must not exceed 200 metres. According to the national definition, only 25 per cent of the Norwegian population lives in scarcely populated areas. Because of the lack of an official definition of *urban* and *rural* areas, it is the definition of *densely* and *scarcely* populated areas that is used to denote the degree of urbanisation (75 per cent). But it is important to note that the term 'densely populated area' is by no means synonymous with the term 'urban area' as used in the rest of Europe (except Nordic countries). What are commonly considered to be rural areas by Norwegian citizens, politicians and planners include a lot of areas officially defined as densely populated areas. The definition of rural Norway as all parts of Norway except the four biggest urban agglomerations (cf. NFR, 1999), accords with the dominant understanding of the term 'rural areas'.

The biggest cities in Norway are much smaller than big cities in more densely populated countries in central and southern Europe. For example, the biggest city in Norway, Oslo, has only around 800,000 citizens (including those in suburban areas), while the next biggest city, Bergen, only has 225,000 citizens. In fact, only four Norwegian cities have more than 100,000 citizens (SSB, 1999). The biggest cities in Norway, therefore, are very modest in terms of population, compared with the big metropolitan areas in central Europe. At the end of the nineteenth century very few towns (i.e. small centres or agglomerations of settlement) existed beside the bigger cities (Hansen and Selstad, 1999). A structure of small and medium-sized centres in rural areas started to develop at the turn of the century, at the same time as the country experienced strong industrial growth. The regional policy in the 1950s and 1960s supported this trend, and by the 1970s this tendency had culminated in a hierarchical structure of urban-type centres.

Hansen and Selstad (1999) suggest a categorisation of Norwegian regions into five types: big city regions (50,000 citizens or more), small city regions (10,000 to 50,000 citizens), town regions (5,000 to 10,000 citizens), village regions (2,000 to 5,000 citizens) and rural regions (200 to 2,000 citizens). According to this classification, 44 per cent of the population lives in big city regions, 30 per cent in small city regions, 13 per cent in town regions, 8.5 per cent in village regions and only 4.5 per cent in rural regions. This shows that most Norwegians do not live in rural areas. In fact, more than half of the Norwegian population lives in or close to the biggest cities of Oslo, Bergen, Trondheim, Stavanger, Kristiansand and Tromsø. However, we should not be fooled by these numbers nor the urbanisation rate of 75 per cent into believing that the Norwegian population is 'urban' in an European sense of the word.

Although contemporary Norway can not be described as strongly urbanised when compared to most other European countries, the country has in the post-war period experienced a relatively strong migration flow from rural to urban areas. In the 1950s, approximately 50 per cent of the population lived in scarcely populated areas. Migration flows to urban areas were modest and with a higher birth rate in rural areas than in urban areas, settlement patterns did not change very much. In the 1960s, however, migration flows to urban areas escalated because of increased industrialisation and strong economic growth. Young people from rural areas moved into the cities where the new jobs, schools and universities were located, and although the rural birth rate still was relatively high, it could not counterbalance out-migration flows. A decline in migration towards the biggest cities and an increase in migration towards rural areas in the early 1970s caused a short period of counter-urbanisation. In fact, only the three biggest cities had a negative net migration rate and a lot of the big city regions actually had a positive net migration rate if we look at the decade as a whole. Migration towards the urban centres increased again in the second half of the 1970s. The 1980s and 1990s can be characterised as 'the decades of urbanisation' (Foss and Selstad, 1997). These decades were marked particularly by strong migration from the scarcely populated areas in the north of Norway, towards the biggest cities in the southern part of the country. There was also a general tendency towards urbanisation in all parts of the country. People left the countryside and moved towards urban centres of all sizes. The population growth in the two last decades was strongest in the big city regions, weaker in the small city regions and weaker still in the town regions. Below this level the population has decreased (Hansen and Selstad, 1999).

There are two main forces behind the urbanisation process in Norway: employment and education (Orderud, 1998). In other words, rural-urban migration is not caused by a widespread wish to live in urban areas. A large proportion of the Norwegian population actually want to live in rural areas. Statistical surveys show that when people are asked where they would prefer to live, a majority express a wish to live in small towns, villages and rural areas (Orderud, 1998). We will return to this when discussing social representations of rurality in the last section.

Regional Economic Restructuring

In the post-war era Norway, like most other West European countries, has experienced strong economic growth. To give a dynamic picture of the economy of post-war Norway in an urban-rural perspective, we address two issues: namely job growth and changes in the distribution of employment in economic sectors. How have the new jobs been distributed between urban and rural areas? How has employment in primary, secondary and tertiary industries respectively been distributed between urban and rural areas?[2]

A large number of new jobs have been created, but they are not evenly spread across the country. The majority of new jobs have been created in the biggest cities, while the most rural areas have experienced a loss of jobs especially during the last two decades. In order to present a more detailed picture of how the growth in the number of jobs has been geographically distributed during the period from 1970 to 1990, we relate job growth to the same categories of regions as used above in the description of the settlement pattern (Foss and Selstad, 1997). The categories 'village region' and 'rural region' are, however, combined in the category 'village and countryside'. Table 11.1 demonstrates this close connection between employment change and population change. The geographical distribution of job change almost

Table 11.1   Geographical distribution of job change in Norway, 1970–90

Type of region (centrality)	Proportion of new jobs Relative change (%)		Proportion of employment
	1970-80	**1980-90**	**1980**
Big city	47	81	49
Small city	30	17	27
Town	12	7	11
Village and countryside	11	-5	13
Norway (total)	100	100	100

Source: Foss and Selstad, 1997.

matches the rate of change in the distribution of population. This is especially evident for the figures from the 1970s which show that both 'village and countryside regions' and 'towns' experienced an 11–12 per cent growth in jobs. The growth in the biggest cities did not stop, however. Almost every second new job was created in the big city regions (Foss and Selstad, 1997). By the 1980s, the situation had changed dramatically. The two categories of city regions now accounted for almost all the job growth, and the big city regions alone accounted for more than 80 per cent. The proportion of job growth in the town regions was now significantly reduced, and in the most rural areas the growth of the 1970s had turned to a decline (Foss and Selstad, 1997).

Norway, together with other developed economies, has witnessed major changes in the composition of economic activity of which the shift in employment structure from extractive to service industries is the most fundamental. As in most advanced industrialised countries about 70 per cent of employment is now in the tertiary sector.

Table 11.2 illustrates the changing importance in terms of employment of the primary, secondary and tertiary sector industries in scarcely and densely populated areas and for Norway as a whole since 1950. It shows that the expansion in service employment was especially strong in the 1970s. This was due to the expansion of the public sector and the high purchasing power in the population in that decade. Analysed from an urban–rural perspective it is interesting to note that the growth in services in the 1970s was almost as strong in the scarcely populated areas as in the densely populated areas. The decade might therefore be termed 'the decade of decentralisation' in terms of its growth in services (Sjøholt, 1983). By 1990, more than half the workforce was employed in service industries across Norway. 'Community, social and

Table 11.2   Distribution of the working population by industry and area type, 1950–90, by percentage

Year		Primary industries	Secondary industries	Tertiary industries	Total
1950	Scarcely populated	53.8	27.7	18.5	100
	Densely populated	1.5	42.8	55.7	100
	Whole country	37.4	32.4	30.2	100
1960	Scarcely populated	29.7	35.3	35.0	100
	Densely populated	1.2	38.8	60.0	100
	Whole country	19.6	36.5	43.9	100
1970	Scarcely populated	32.6	33.8	33.6	100
	Densely populated	1.7	44.1	54.2	100
	Whole country	11.0	41.0	48.0	100
1980	Scarcely populated	25.1	29.7	45.2	100
	Densely populated	1.7	29.6	68.7	100
	Whole country	8.1	29.6	62.3	100
1990	Scarcely populated	18.0	27.5	54.5	100
	Densely populated	1.6	24.9	73.5	100
	Whole country	6.1	25.6	68.3	100

Source: SSB, 1950, 1960, 1970, 1980, 1990.

personal services' was the category showing highest growth in scarcely populated areas. Second came 'Financial, insurance, real estate and business services' and third 'Wholesale and retail trade' (Teigen, 1999). A relatively detailed study of changes in service location in the region of Middle-Norway between the mid-1960s and the mid-1980s using a threefold categorisation of services concluded that 'Information functions' was the most rapidly expanding category in the region as a whole, but had been the most expansive in the middle tier of the central place system. In the region as a whole the growth in 'Reproduction functions' was slightly bigger than the growth in 'Distribution functions'. Private reproduction functions had been most expansive in the middle tier, while the public ones had expanded the most in the lower tier. The distribution functions had expanded in the middle and upper tiers of the central place hierarchy (Berg and Sjøholt, 1988).

The proportion of the Norwegian working population employed in secondary industries grew until 1970 when it reached 41 per cent. Since the mid-1970s the proportion has steadily decreased and today is about 25 per cent. This 'deindustrialisation' has been more severe in densely populated areas than in scarcely populated areas, a process which has led to the fact that secondary industries employed between 25–30 per cent of the workforce in both densely and scarcely populated areas in the 1980s as well as the 1990s.

As Table 11.2 shows, employment in the primary industries has been dramatically reduced in the post-war period. Although the sector employed more than half of the workforce in 1950, it encompassed only 6 per cent of the working population in 1990. Since the percentage in primary employment in densely populated areas has been marginal and relatively stable in the whole period, the reduction has taken place mainly in scarcely populated areas. This has of course played a significant role in the process of urbanisation. Of the 6 per cent employed in primary industries, 5 per cent were employed in farming, the remainder in forestry, fishing and hunting (SSB, 1990).

Figure 11.1 shows the pattern of sectoral restructuring in scarcely populated areas in the post-war period. It illustrates clearly the rapid growth of the tertiary sector since 1970 and demonstrates that it is by far the largest source of employment in rural areas. Although the share of employment in primary industries is only 18 per cent in scarcely populated areas (1990), some argue very strongly that primary employment – and first and foremost farming – is important for rural employment. For instance, Blekesaune (1999) in an analysis of the importance of rural farming for rural employment employs a rurality index developed by Almås (1995) which categorises the 435 Norwegian local authority areas as rural (171), semi-rural (100), semi-urban (81) and urban

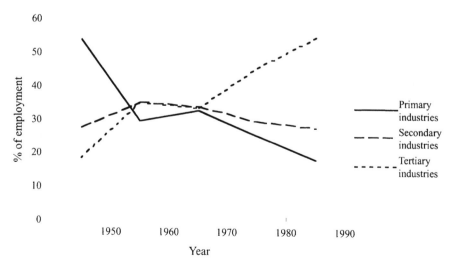

Figure 11.1 Changes in working population by industry, scarcely populated areas, 1950–90

(83). According to this classification, the farm sector employs more than one sixth of the labour force in rural local authority areas, and if rural and semi-rural local authority areas are combined in one category the share is one quarter. Blekesaune (1999, p. 27) concludes that 'even if employment in farming has decreased during the last decades, farming still represents a substantial part of rural employment'. To modify the statement somewhat we would like to emphasise that one of the indicators used to calculate the rurality index is the proportion of the workforce in primary industries.

Nonetheless, the decreasing share of employment in primary industries in rural areas illustrates a change that has been described as a shift from agricultural productivism to post-productivism (Ilbery and Bowler, 1998), a change evident in Norway as in most advanced economies. Rural household members take part in the generation of income from on-farm and/or off-farm sources in addition to income obtained from primary agriculture. Examples of on-farm sources are farm-tourism, especially farm-based accommodation and recreational activities, while off-farm sources include income from many types of employment outside agriculture, not least in service industries, and to a certain degree also self-employment. This pluriactivity, which undoubtedly is a survival strategy adopted by farm households, makes farm households very important for maintaining the supply of different types of service in rural areas as members of these households represent both service demand and work force (Aasbrenn, 1998).

The dynamic and rapidly growing Norwegian service sector has actually brought about a new economic geography, in that different parts of Norway have become more like each other. A hallmark for Norway has been a national deconcentration of services, especially public services. This is not to say that there are no important differences between urban and rural areas. The development of rural Norway, in spite of the growing service sector, is characterised by the loss of jobs and the loss of labour (especially skilled labour) in favour of more urban areas, and the share of the workforce employed in services is 19 percentage points higher in urban than rural areas.

Norwegian Administrative Structures and the Distribution of Power

This section gives a short overview of the administrative structures governing the management of regional development in Norway. Our aim is simply to provide an understanding of the significance of central and local levels of decision-making.

There are three administrative levels involved in regional development in Norway: central government, the 19 counties (*fylker*) and 435 local authority areas (*kommuner*). Regionally elected local governments run the administration of the counties and the local authority areas. Because regional policy in Norway until relatively recently had been a policy for the redistribution of resources towards priority regions (see next section), regional policy had been administered by central government. Government at the local levels had mostly been concerned with the administration of public services within rather narrow legal and financial limits (Mønnesland, 1997). A decentralisation process accelerated in the 1980s caused by factors including the relatively favourable development of rural areas during the 1970s (as described above) which had resulted in regional self-confidence and the new problems that the state-run regional policy encountered as unemployment grew in urban areas. In this situation it was, as Mønnesland (1997, p. 29) states, 'convenient for the central authorities to delegate more responsibility to lower regional levels'. It was therefore stated more or less explicitly by the central authorities that regional development ought to be the responsibility of each region.

What really happened was that given the transfer of resources, regional authorities gained responsibilities for the design of suitable measures for regional development (Mønnesland, 1997). The philosophy behind this move was that decisions made at the county or municipal level would probably be most responsive to regional or local situations. The counties have gained a strong position in many respects, not least because of the broad responsibilities assumed by them, i.e. transportation, health care, education, regional planning and especially issues of economic development (Lindstrøm, 1996).

Although the counties have improved their position in relation to state authorities, they still lack power in order to shape and implement regional policies. There have been problems due to this imbalance of political power at different political-administrative levels, and these have been expressed in a study of cooperation across nation state borders between regional governments (counties) in Norway, Sweden and Finland in the middle of Scandinavia – the Mid-Nordic Cooperation (Lysgård, 1997, 1999, forthcoming). This cooperation is based on a wish to solve problems caused by the relatively small population of the area, and on a wish to develop an alternative to the strong attraction on both finance and skilled labour exerted by the capital cities of Oslo, Stockholm and Helsinki. This project of making and implementing regional policies at a decentralised level as an alternative to state policies has not been very successful. The reason for this is mainly related to problems concerning the strong power held at the national level when it comes to decision-making.

For example, much effort has been made to coordinate planning in the development of infrastructure for transport throughout the area. However, problems have arisen in the implementation of plans because decisions have to be made at the national level where neither the plans nor the cooperation as such figure as national priorities. In other words, there is a top-down negligence of the idea of regional cooperation as such. In addition this is also opposed from below, evident in the low levels of interest from local and regional populations (Lysgård, forthcoming).

The local authorities' position is, however, very much dependent on the agency of individual actors or group of actors both within and outside the administrative or political bodies. This in turn creates, on the one hand, pro-active local authority areas that choose reformation strategies, in which they aim to reform and adjust the local authority area to the new general framework imposed by national and international structures, and, on the other hand, reactive local authority areas that choose strategies of resignation and 'muddling through', or even resistance strategies that imply protest against any types of changes and/or calls for the national authorities to compensate their losses. Ultimately even the power of pro-active local authority areas to influence their own development is very much decided by the national authorities and their existing and potential comparative advantages (Halvorsen, 1996).

Regional Policy

Having summarised very briefly the main trends in regional development in Norway since the Second World War and described the administrative structures governing regional development, we now turn to consider Norwegian regional policy. In the 1950s, regional development was defined as a special field of politics aimed explicitly at influencing regional development. Local affiliation is very strong in Norway probably as a consequence of the natural landscape and the small population. Fjords and high mountains have resulted in relatively small and isolated local communities that in turn has resulted in a lack of urban traditions. This is probably also the reason why regional priorities have had a reasonably high profile in other areas of politics in the period. Our aim here is restricted to providing a survey of explicit Norwegian regional policies, that is, incentives administered by regional policy authorities. We focus on changing philosophies and goals, rather than measures, although these of course are closely connected.

Norway and the other Nordic countries, except Denmark, have vast areas with extremely low population densities. These areas have been the focus of Nordic regional policy in most of the post-war period until relatively recently. Nordic regional policy in the post-war period has, therefore, been what Mønnesland (1997) in English calls 'periphery policy'. But as he also writes that:

> The differences compared to EU traditions are illustrated by the fact that 'periphery policy' has no clear English translation. The normally selected English term is 'regional policy' both for general regional policy and the more specific 'periphery policy' (p. 16).

Mønnesland (1997) does not discuss the term 'rural policy' which is another alternative. Although 'urban' and 'rural' have different connotations in Norway than in central and southern Europe, as we have discussed above, it is as relevant (if not more so) to talk about 'rural policy' as 'regional policy' because of the rural focus of it. Consequently we use both terms.

In this section we describe the regional policy in the 1950s, 1960s, 1970s, 1980s and 1990s respectively. After the Second World War, Norway was faced with the task of rebuilding the country. There was economic growth and an increase in resources available for distribution. It was almost undisputed that the benefits of economic growth should be regionally distributed, and that a regional policy was needed for this purpose. 'Town and country – hand in hand' was a widely accepted slogan in the 1950s (Teigen, 1999). Regional balance was set up as a goal, and less developed regions had to benefit. Northern Norway in particular was regarded as a less developed region, and development programmes aiming at economic growth in the region were initiated. The development of huge state-owned manufacturing plants was an important measure in that respect. Manufacturing was in general considered to be the industry for the future and industrialisation of rural areas was seen as a process which would hamper tendencies towards economic centralisation.

In the 1960s the period of the construction of state-owned manufacturing plants was over. The aim of creating a foundation for further industrialisation was regarded as largely achieved. The decade was characterised first and foremost by the establishment of a nationwide bureaucracy to administer the regional policy. Attention turned also to regions other than the northern parts of Norway. The rationalisation processes in agriculture caused unemployment in most rural areas which in turn caused migration to urban areas and consequent urban growth. Norway's rural areas faced severe problems. Regional policy discriminated on the basis of negative selection, i.e. regional policy was meant to stimulate economic growth outside defined congested areas.

In the 1970s, the idea of even economic growth in all parts of the country was consolidated with the additional idea of stabilising settlement patterns. This idea has permeated Norwegian regional policy ever since. The goal was well on the way to achievement during the 1970s, since there was, as described above, a decrease in net out-migration from peripheral regions and a decrease in in-migration to central regions. In other words there had been a relative stabilisation of the pattern of settlement. An important explanation to this trend is the strong growth in employment in the public sector that took place particularly in rural areas during this decade. Another is that at the same time, i.e. from 1974 onwards, manufacturing industries in the urban regions experienced a stagnation, partly due to laws restricting investments in urban areas, while employment in manufacturing continued to grow in the rural regions, partly as a result of the development of manufacturing parks by local authorities, transportation subsidies, reduced employer taxation, cheap loans and investment support (Teigen, 1999).

A response to these new trends in regional development during the 1970s was a reassessment of the regional policy in the late 1970s and early 1980s (Mønnesland, 1997). In the early 1980s manufacturing industries in peripheral regions which had been given priority in regional policy began to stagnate and the development of public services in these areas had been achieved. This meant that the main forces behind the relative stabilisation of the settlement pattern in the 1970s had ceased to be so important. The result was a new wave of migration to central regions, caused mainly by employment opportunities. The objective of stabilising the settlement pattern became as important and as difficult to achieve as it had been in previous decades. In addition to considering migration to the main population centres as a problem for the rural regions, as had been the case in the 1960s, attention also turned to emerging problems in the big cities. Urban problems had entered the policy agenda.

The late 1980s, however, saw a decline in the tendencies towards stagnation and employment problems in most parts of the country and the problem of rapid expansion (including the Norwegian capital) in urban areas. Whole regions with a cluster of involved firms experienced structural problems as a result of industrial recession (Mønnesland, 1997). Again a reassessment of the regional policy, not least its geographical profile, was needed.

> Problems of this nature have partly led to a redefinition of the geographical profile of regional policy, and partly to a changing focus from peripheral oriented

incentives towards structural and nation-wide industrial policy measures. They have also led to a new debate on the legitimacy of regional policy (Mønnesland 1997, p. 26).

In this situation, with new nationwide problems, rather than specific rural or urban problems more responsibility was delegated to lower regional levels and attention was turned to local entrepreneurship rather than redistribution as a solution. Among other things, a programme for the development of private service firms was established to contribute both to employment and service supply.

The Review to Parliament by the Minister of Regional Policies in 1992–93 marks a turning point away from rural (periphery) policy to regional policy (Teigen, 1999). In the early 1990s, increased internationalisation of trade was brought up as an argument against favouring firms in rural areas, since the firms capable of competing internationally are usually located in the biggest urban areas. As a non-EU member, Norway is not a recipient of the Structural Funds,[3] but as a member of the European Free Trade Association (EFTA) Norway has from 1994 been a signatory for the EEA (European Economic Area) Treaty. According to Mønnesland (1997), the EEA rules have led to some reductions in the support given to enterprises, but smaller firms have mostly avoided much of the reductions.

While the first half of the 1990s can be characterised as a period of relative stability as regards settlement patterns, the second half of the decade saw an alarming increase in migration from peripheral regions. A big difference compared with the beginning of the decade was that larger centres in the more remote regions that used to balance the total figure of the region, were increasingly faced with migration loss (Hanell, 1998). Examples are cities such as Tromsø, Bodø, Mo i Rana and Narvik in Northern Norway. Towards the end of the 1990s there was therefore, again, a shift in the focus of regional policy. Aalbu (1998) describes the Review to Parliament by the Minister of Regional Policies (in March 1998) as '... a shift from a policy directed towards all regions towards a policy with a stronger focus on the weakest parts of the country' (p. 9). The shift can also be considered a consequence of a change of government from Labour to a three-party government in which the Agrarian Party was an integral part. The Agrarian Party has traditionally stressed rural values and promoted the belief in the ability to actively influence regional development. It is, in other words, both the new migration trends and the Agrarian Party's attitude that underlie the change of policy. In the Review to Parliament it is stated that municipalities should aim to maintain their number of inhabitants and the measures suggested are better coordination of public policies, such as the introduction of a system of regional impact analyses in all sector ministries, stronger financial efforts in the support zones directed towards young people, increased efforts in education and research, IT and SMEs and promoting the image of rural areas. The latter measure is new in Norwegian regional policy and therefore especially interesting. It is interesting also because rural studies conducted in other countries during the 1990s have shown that images of the rural seem to lie behind migration into rural areas (see for example Boyle and Halfacree, 1998). In the next section, we turn therefore to images, or social representations, of Norwegian rurality and urbanity and suggest that they tie into processes of regional development, especially migration flows. It is important to note at this stage that the Labour Party after a cabinet crisis in March 2000 is once more in power and that a new Review to Parliament from the Minister of Regional Policies is expected by 2001. It remains to be seen whether the focus on the weakest parts of the country will be continued.

Social Representations of Rurality and Urbanity

There are, as we see it, two dominant social representations of rurality and urbanity respectively within contemporary Norwegian society. First we have the representation of rural areas as beautiful, safe, clean, harmonious, peaceful, and healthy, where everyone knows each other and takes care of each other (a type of rural idyll) versus the ugly, dangerous, polluted, unhealthy and noisy city with a lot of lonely people and no neighbourliness. Second there is the representation of rural areas as traditional, backward, boring places with only one grocery store, one post-office, one petrol station and one cafe (written 'kafe' with *k* as the Norwegian dictionary tells) versus the modern, exiting city with everything one needs, including theatres, galleries, musical events and cafes (written 'cafe' with *c*) available.

The significance of social representations for migration flows has not been addressed significantly in Norwegian regional policy discourses until quite recently, nor in mainstream academic discourses on migration flows. Taking as the point of departure the finding that most Norwegians prefer to live in rural areas, villages or small towns but are pushed towards urban areas by the need for employment and education (Orderud, 1998), it is interesting to ask: why is it that people find rural areas so attractive? How do they conceive of rural areas? What are people's everyday interpretations of the concept of rurality and the places they see as rural? Inspired by moves in the British rural studies literature towards considerations of the social and cultural construction of the rural and recognition of the constitutive power of lay discourses (see, for example, Halfacree, 1993, 1995; Jones, 1995; Valentine, 1997) an ongoing study focuses on exactly these questions (Berg, 1998). It takes as its point of departure the fact that although the majority of migrants move from rural to urban areas, there are some who move in the opposite direction. Could the same approach be employed as in the British studies in order to understand counter-urbanisation, and could this be fruitful in an analysis of the small-scale urban-rural migration in Norway? Does the representation of the rural as 'idyll' also play an important role in Norway? Using life-history interviews with 22 in-migrants (11 couples) to rural areas the study aims at highlighting the significance of social representations of rurality for the respondents' own constructions of the rural while living in the city and their decision to migrate to a rural locality as well as their continuous reconstructions of and negotiations about the meaning of the rural after having settled in a rural locality.

There is no space for a comprehensive summary of the conclusions from the study within this chapter, but some of the conclusions about the main elements of the urban-rural migrators' constructions of rurality before moving should be mentioned. First, rural areas are constructed as idyllic settings for family life, not least because they are thought of as safe places for children to grow up in, as well as being understood as offering more possibilities for environmental exploration and imaginative play. Second the outdoor life, hunting and fishing are important elements in the respondents' constructions of the rural. While outdoor life is conceived of as a positive element in rural life by all the interviewees, hunting (and fishing) are thought of as gendered activities. While women who are negative towards hunting see the activity as time-consuming and therefore a threat to family life, those men who consider hunting as a negative element see it as an ingredient of a hegemonic rural masculinity with which they feel uncomfortable. Most of the men and a couple of women taking part in hunting see it as a healthy, exciting and useful activity that is a traditional and an obvious or inevitable part of rural life. Third, and partly interlinked with hunting, the interviewees conceive of the rural as gendered; that is, they see constructions of rurality as intertwined with constructions of femininity and masculinity and containing rules about appropriate or legitimate behaviour and actions by men and women both in relation to types of economic activity and hobby activities that are more traditional or conservative than in urban areas. Most of the respondents characterise rural areas as masculine, as areas where male activities and norms in general are the most visible and valued. In studies on rural-urban migration and particularly in the media, the 'hairy-chested' character of rural areas has been put forward as an explanation as to why young women leave the countryside and do not return (Dahlstrøm, 1996). The study indicates that the 'masculine' countryside also makes some men sceptical towards moving from urban to rural areas. Finally, the interviewees construct rural life as relaxed and less complicated and crowded than urban life with, for example, better access to basic services such as dentists, doctors, post-offices, hairdressers and kindergartens because of no or short waiting times or queues.

Two main conclusions can be drawn. First, the interviewees' constructions of the rural while still living in the city contained diverse and even contradictory elements. There are elements of the social representation of the rural as idyll, but there is no naïve acceptance of the idyll. The representation of rural areas as traditional and backward, as regards gender relations, for example, was also present. Second, social representations of rurality had played a significant role in the decision-making process. The study did not aim to analyse questions concerning the origins of representations of the rural (and the urban) and the identity of the image makers. The interviews leave no doubt however, that television programs, newspapers, journals, films and literature contribute to the constructions of rurality held by potential in-migrators' to rural areas. Future research should undoubtedly face the challenge of studying the image-makers, the images and their power to influence regional development.

Conclusions

Although the main trend in Norway is rural-urban migration, as opposed to trends such as urban-rural migration in Britain, the approach to migration that is based on an understanding of the rural and the urban as social representations seems to be able to contribute to a more nuanced understanding of people's migration decisions and the resulting migration flows in Norway. The existence and character of different ideas of the rural and how these ideas are powerful in connection with migration, as has been stressed above, are issues that need to be addressed more thoroughly. This is a challenge to future research on regional development.

As regards future regional policy, we observe some changes concerning how rural places are conceived of as policy target areas in official writings as well as in the media; there has been a shift from viewing them mainly as labour markets to viewing them also as dwelling places. It is suggested that questions about how rural areas' attraction might increase should be included within rural policy discourses, not only as labour markets but also as dwelling places. It is in line with this type of reasoning that 'improving the image of rural areas' has recently been suggested as a policy measure by the ministry of local and regional affairs (KRD, 2000).

Notes

1 The definition was originally suggested by the geographer Hallstein Myklebost (1960) and included in addition to the physical criteria of today also a functional criterion, which is that at least three-quarters of the economically active population should be working within secondary and tertiary occupations. According to Hansen (2000), Myklebost's definition was used in the 1960 and 1970 population censuses, but in 1980 Statistics Norway did not have the resources to collect and analyse data related to the functional criterion.

2 This analysis is based on statistics from Norway's Population and Housing Censuses which are conducted every tenth year. Because the 2000 census is still not finished, it is impossible to present data from after 1990. There are available data on economic activities in the 1990s, but these are not disaggregated between densely and scarcely populated areas, which are the main points of interest here.

3 The only EU regional policy programme in which Norway participates is the Interreg programme for cooperation between regions across established internal and/or external borders of the EU. In 1995 several such cooperative agreements were established between regions in Norway, Sweden, Finland and Russia, more or less based on the former established system of Nordic cross-border cooperations. Ironically, although Norway had said 'no' to participation in the European Union and its common policies in 1994, it actively supports the Interreg programme, without being able to influence the policies' goals and strategies. Without being able to get any financial funding from the Union, Norway actually put more money into the Interreg programme at the time than, for example, Denmark, which has been a member of the EU since the early 1970s (Hedegaard, 1996).

References

Aalbu, H. (1998), 'The Pendulum of Regional Policies in Norway and Sweden', *North*, vol. 9, pp. 7–10.

Aasbrenn, K. (1998), *Regional Tjenesteorganisering i Uttynningsområder. Utfordringer – Erfaringer – Strategier*, Report No. 5, Høgskolen i Hedmark, Landbruksforlaget, Oslo.

Almås, R. (1995), *Bygdeutvikling*, Det Norske Samlaget, Oslo.

Berg, N.G. (1998), *Kjerringer og Gubber Mot (Flytte)strømmen – Hvorfor Noen Flytter til Bygde-Norge*, lecture held in Tydal, 4 May 1998.

Berg, N.G. and Sjoholt, P. (1988), 'The Role of Commercial and Other Service Functions in Changing the Structure of the Central Place Hierarchy', *Norsk Geografisk Tidsskrift/ Norwegian Journal of Geography*, vol. 42, pp. 1–12.

Blekesaune, A. (1999), *Agriculture's Importance for the Viability of Rural Norway*, Report no. 8/99, Centre for Rural Research, Trondheim.

Boyle, P. and Halfacree, K. (eds) (1998), *Migration into Rural Areas: Theories and Issues*, John Wiley & Sons, London.

Dahlström, M. (1996), 'Young Women in a Male Periphery – Experiences from the Scandinavian North', *Journal of Rural Studies*, vol. 12, pp. 259–71.

Foss, O. and Selstad, T. (1997), *Regional Arbeidsdeling*, Tano, Oslo.

Halfacree, K. (1993), 'Locality and Social Representation: Space, Discourse and Alternative Definitions of the Rural', *Journal of Rural Studies*, vol. 9, pp. 23–37.

Halfacree, K. (1995), 'Talking about Rurality: Social Representations of the Rural as Expressed by Residents of Six English Parishes', *Journal of Rural Studies*, vol. 11, pp. 1–20.

Halvorsen, K. (1996), 'Hvor Stort Albuerom har den Norske Distriktskommunen?', in K. Aasbrenn (ed.), *Opp og Stå, Gamle Norge*, Landbruksforlaget, Oslo, pp. 32–45.

Hanell, T. (1998), 'Nordic Peripheries in Trouble Again', *North*, vol. 9, pp. 11–15.

Hansen, J.C. (2000), 'Hallstein Myklbost 1923–2000 – an Obituary', *Norsk Geografisk Tidsskrift/Norwegian Journal of Geography*, vol. 54, in press.

Hansen, J.C. and Selstad, T. (1999), *Regional Omstilling – Strukturbestemt eller Styrbar?*, Universitetsforlaget, Oslo.

Hedegaard, L. (1996), 'Interreg i Norden – Geopolitik eller Samhørighed', *Nordrevy*, vol. 96, pp. 10–18.

Ilbery, B. and Bowler, I. (1998), 'From Agricultural Productivism to Post-Productivism', in B. Ilbury (ed.), *The Geography of Rural Change*, Longman, London, pp. 57–84.

Jones, O. (1995), 'Lay Discourses of the Rural: Developments and Implications for Rural Studies', *Journal of Rural Studies*, vol. 11, pp. 35–49.

KRD (Kommunal og Regional Departement) (2000), *Distrikts-Norge – Hvor Ligger Mulighetene – Hva Må Gjøres*, Memo for discussion about future regional policy in Norway, Kommunal- og regionaldepartementet: http://odin.dep.no/krd/.

Lindström, B. (1996), *Regional Policy and Territorial Supremacy*, NordREFO, 1996, 2.

Lysgård, H.K. (1997), 'To Be or Not to Be a Region – Is that the Mid-Nordic Question?', *Nordisk Samhällsgeografisk Tidskrift*, vol. 25, pp. 68–79.

Lysgård, H.K. (1999), 'Produksjon av en Region. Et Eksempel På Hvordan vi Former, Oppfatter og Identifiserer oss med Romlighet – Midt-Norden Som Case', in. S. Helmfrid (ed.), *Regionala Samband och Cesurer*, Tapir, Trondheim, pp. 9–22.

Lysgård, H.K. (forthcoming), *Produksjon av Rom og Identitet – Et Eksempel fra det Politiske Samarbeidet i Midt-Norden*, Doctoral thesis, Department of Geography, NTNU Trondheim.

Myklebost, H. (1960), *Worges Tettbygde Steder 1875–1950*, Universitetsforlaget, Oslo.

Mønnesland, J. (1997), *Regional Policy in the Nordic Countries. Background and Tendencies 1997*, NordREFO 1997, 2.

NFR: Norges Forskningsråd (1999), *Regional Utvikling*, The Norwegian Research Council's Research Programme for Regional Development 1999–2003.

Orderud, G.I. (1998), *Flytting – Mønstre og Årsaker. En Kunnskapsoversikt*, NIBR prosjektrapport 1998, 6.

Sjøholt, P. (1983), '1970-årene – Tiåret for Desentralisering av Serviceaktivitetene i Norge?', *Plan og Arbeid*, vol. 3, pp. 163–71.

SSB: Statistisk sentralbyrå (1950–90), *Population and Housing Census*.

SSB: Statistisk sentralbyrå (1999), *Statistical Yearbook 1999*, Statistics Norway, vol. 118.

Teigen, H. (1999), *Regional Økonomi og Politikk*, Universitetsforlaget, Oslo.

Valentine, G. (1997), 'A Safe Place to Grow Up? Parenting, Perceptions of Children's Safety and the Rural Idyll', *Journal of Rural Studies*, vol. 13, pp. 137–48.

Chapter Twelve

Conclusions

Keith Halfacree, Imre Kovách and Rachel Woodward

Acknowledging Continued Diversity

> Where we still fall short in rural studies is in analyses of structuration processes. ... Today, we fear, too much research is falling between two desirable goals; quality research of local level processes and theoretical contextualization of the structured circumstances of activities in rural areas. ... For theoretical advance we need at least to debate and explore the role of nation-specific forces and not take them for granted, or even worse ignore them (Hoggart et al., 1995, p. 264).

> Rural development seems, in many important respects, to have a life of its own. Despite concerted efforts by both state agencies and private-sector firms to discover a secret recipe for economic success in the countryside, rural areas continue to follow their own stubborn logics of change and stasis (Murdoch, 2000, p. 407).

The remit of this edited collection was straightforward: to examine a set of questions concerning the context for and operation of rural development practices across different European countries (see Chapter One). It thus sought to follow, to some extent at least, the call made in the quotations above. This remit was conceived and evolved at a time of considerable change, critical questioning and even soul-searching within Europe generally. More specifically, it was felt by the editors that in a Europe characterised within dominant discourses of economics, politics, culture, society, and geography as experiencing convergence above all else, there was a clear need to evaluate this experience as it was played out on the ground within the rural areas of each country. The primary focus would be a ground-clearing exercise to obtain a comparative sense of both the geographies of rural Europe and the processes responsible for stasis and change, with the additional intention of making special reference throughout to policies and practices of development. In summary, we may ask to what extent are we now able to talk meaningfully of a 'rural Europe' in the context of *Leadership and Local Power in European Rural Development*. Or is heterogeneity still the key emergent feature of the rural areas of Europe?

The collection has gone some way to addressing this central question. Overall, in spite of a number of general themes and common experiences, not least trends encompassed within a (highly uneven) shift from production towards consumption across much of rural Europe, there is still great diversity concerning the issues implicated within rural development (see also Hoggart et al., 1995). Consequently, policies originating within the EU's Common Agricultural Policy and Structural Funds, for example, must include a space for this diversity at all stages, from conception through to implementation and evaluation. Whatever the merits of universalised discourses of 'globalisation', rural Europe remains an excellent setting in which to explore and evaluate the rhetoric and realities of these discourses. In spite of the need to recognise the key structuring role played by national and even global processes, rural Europe is still characterised as having a 'rural mosaic' (Hoggart et al., 1995, pp. 21ff).

Within this collection, diversity came across at all levels of investigation. However, whilst on the one hand this reflected the characteristics of the different countries surveyed, on the other it also reflected the researchers responsible for each chapter. The contributors were a heterogeneous group of scholars in terms of their disciplinary affiliations (geography, sociology, political science, gender studies) and the conceptual approaches they took towards the subject. From such a group, the responses to the remit of this book, as set out in the preceding chapters, was inevitably diverse. The similarities and differences in approach to the set of questions framing the collection, and the comparisons and contrasts which can be drawn between different national contexts, will be obvious enough through a review of the text.

The diversity shown in this collection does not mean, however, that it is impossible to draw more general lessons for rural development. This is important to stress since, as we argued in our introductory chapter, the rural development issues discussed in the collection are significant ones for the overall economic prosperity and social cohesion of the countries concerned. European rural development policies and programmes need to learn from experiences of implementation in different national contexts, just as they need to be based on a rigorous, theoretically-informed conceptualisation of the logic and consequences of development policy. This volume, with its detailed exposition of national contexts and specificities, stands as a contribution to that process. There are clear lessons to be learnt from the diversity of political configurations, European policy experiences, regionalist tendencies and constructions of ruralities outlined in the preceding chapters. Our purpose in the remainder of this conclusion is not, however, to reiterate these lessons.

This is not least because we feel that they need always to be understood within the context through which they arise, which makes them hard to express boldly and succinctly. Instead, drawing on some of our own lessons as editors, we wish to single out three fruitful avenues for future comparative research.

Areas for Future Research

National Contexts and Definitional Issues

Relationships between national contexts and the concepts deployed in the analysis process raise some interesting questions for comparative research on European rural development. This comparative project started with a template of basic issues to be addressed in the search for common ground. Apart from differences stemming from contrasting disciplinary backgrounds and traditions, as noted above, there were also issues about both the understanding and the applicability of some of the elements of the original template within different national contexts. Some expressions of this were immediately obvious. For instance, what sense did it make to talk about the efficacy of EU development policies for Hungary or Norway, both non-EU members? Other expressions of difference were more ingrained within contrasting national or disciplinary cultures.

An excellent example of this definitional indeterminacy concerns the applicability of the concepts of 'rurality' and 'the rural'. In Germany, for example, the authors argue that the concept has little utility in understanding power and rural development in the reunified country. This is because they consider rurality in Germany as being a 'secondary concept', usually subordinate to ideas such as 'region', 'peasant' or 'periphery'. In particular, they argue for most attention to be paid to the significance of regionalist discourses in understanding national approaches to regional disparities, and regional responses to common development problems. Given such an emphasis, what does this mean for the concept of rurality and its applicability in further European comparative work? And what of the conceptualisation of regionalism? Might the latter come to have greater resonance within development issues emerging in strongly centralised states, such as the UK, where regional governance has only recently emerged as a viable administrative tool? A further example would be the conceptualisation of the roles of elites in different national policy contexts, and specifically, their role and power in shaping EU rural development policy; see, for example, the Hungarian example and lessons this holds for other European contexts facing debates over the redistribution of resources. Ultimately, therefore, different national intellectual traditions in combination with different national policy discourses in turn combined with different ways of thinking about rural development needs to produce very different approaches to the key questions guiding this book.

Different countries' contrasting state structures are also critical to understanding the extent to which even seemingly transnational rural development policies – notably those arising from within the EU – can come to embody very different meanings and nuances. This is because, in spite of talk of the 'hollowing out' of the nation-state (Jessop, 1994) both upwards (especially to the EU) and downwards to the grassroots, this tendency is still highly uneven across Europe. For example, in France we have a dualist political structure based on a multitude of small municipalities (e.g. rural communes), on the one hand, and a strong central state (e.g. the agricultural administration), on the other. This polarity comes across in what the French contributors labelled a bottom-up/top-down conjunction for rural development. In contrast, Germany is well known for the strength of its regional government, the Bundesländer, with the principle of subsidiarity reflected in the national constitution. Thus, rural development policies come to be reflected through a very different lens than in France, for example, notwithstanding the role of the federal state in striving for 'comparable living conditions' across its constituent parts. There are also comparison to be drawn, for example, between national responses to EU rural development programmes and changes in administrative structures in different countries; see, for example, the experiences of PRODER in Spain and POMO in Finland. Given the significance of such geographical differences to all elements of the policy process, it is vital yet again to stress the constructed nature of the political environment within any analysis. Such a concern has recently been reflected more broadly in Goodwin's (1998) call for more attention to be paid to the governance of rural areas, that is, to all aspects of and actors involved in the government of rural places.

By arguing for an appreciation of the importance of national contexts and associated definitional understandings, with their contributions to forming, *inter alia*, conceptual approaches, we must stress that we are not arguing for a deterministic relationship between development outcomes and circumstance. Similarly, one cannot 'read off' intellectual praxis from these contrasting contexts either. Nor do contexts themselves determine the theoretical approach to be adopted. Rather, our point is to argue for an explicit element of reflexivity in our collective attempts to derive broad lessons from common rural development experiences across diverse European contexts. More specifically, there also seems a necessity for more comparative research into the meanings of terms such as 'rurality', 'elites' and 'the state', instead of assuming that they have a strong degree of equivalence across the work of academics from different European countries.

Tracking Concrete Outcomes

The issue of definitional indeterminacy brings us on to our second key area for future research within the context of *Leadership and Local Power in European Rural Development*. This relates to the simple point that even with some degree of control over our terminology, the development policies themselves remain indeterminate concerning the concrete changes actually occurring within rural places. The chapters, with their examination of the economic, sociocultural and political dynamics of rural areas, all suggest a tricky path taken by rural development, strewn all along with obstacles and false trails. Thus, as is intended by policies such as LEADER, whilst in general even small-scale initiatives can kick into life a dormant or previously nonexistent development agenda within rural Europe, its subsequent passage can be highly problematic. Again, the role of local elites and the local manifestation of power relations is critical here. Policies can be captured, derailed or even blocked completely according to these local power networks.

This issue of recognising the contours of the practical deployment of 'development' is reflected in concerns in the UK that urban interests may be drowning out rural problems within the new Regional Development Agencies. It also comes across from the seemingly highly ingrained power relations within rural Ireland, and in the partial capture of the development process by the 'alternative movement' in rural Germany. Within many of these cases, the importance of migration trends in restructuring the rural population is also clear, with consequent effects upon the practical deployment of development policies.

Recognising the indeterminacy when 'abstract' policy is 'immersed' at the local level (Blomley, 1994) requires a future research agenda to be richer with respect to longitudinal material than has been the case to date. Cross-sectional material is important, of course, but 'rural development' also needs to be tracked from at least just prior to the start of the formulation process to well after programmes and policies have been implemented (e.g. Ellis, forthcoming). A research methodology such as that advocated by Actor Network Theory seems highly appropriate in this respect, with its emphasis on following 'the actors through the network as they build and shape its contours' (Murdoch, 1995, p. 753). Indeed, Murdoch (2000) has called for 'networks' to become 'a new paradigm of rural development', transcending a dualist perspective which sees development as coming largely from either without (top-down/exogenous) or within (bottom-up/endogenous).

Teasing out the networks of development through Actor Network Theory will draw attention to the relative success of different actors – not least those who become the rural elites – in enrolling others into their networks, whilst also 'attend[ing] to the ways that new network forms interact with pre-existing socioeconomic structures' (Murdoch, 2000, p. 416). Moreover, from an Actor Network perspective, the power to control and shape rural development is far from preordained but emerges from the network: 'those who are powerful are ... those able to enrol, convince, and enlist others into networks on terms which allow [them] to 'represent' [speak for] the others' (Murdoch, 1995, p. 748). An open-ended, more ethnographic style of research seems essential for such a task.

Gender and Power Relations

Concern with power relations operative within the rural context takes us to our final area for future work. This involves teasing out the central position of gender within *Leadership and Local Power in European Rural Development*. Whilst gender represents just one form through which power differentials are represented and expressed, its crucial importance in a rural context is clear throughout this collection. A key concept here is power and its forms at the intersection between gender and rurality.[1] A key focus might be the ways in which gender relations operate within and shape the outcomes of the rural development process. There are issues of power and gender in action, concerning the practical and policy issues which follow from questions of gender and rural development. For example, with the decentralisation of the administration of rural development comes questions about control (power over) rural development. What power is kept at national levels? What is transferred? Some of the chapters in this collection have indicated that this process is gendered; what are the nuances of this process, and how can observations across different European contexts be conceptualised? There are also issues of power and gender in ideology. As some contributions to this volume have noted, discourses of rurality abound in the theorisation and practice of rural development. Discourses are a resource, mobilized according to the interests of different actors and groups. How, therefore, do different discourses of rurality construct and reproduce discursive constructions of gender in rural areas?

Conclusion

In summary, having helped to clear the ground, this collection also suggests a range of key issues still to be addressed. Three related areas have been drawn out here. Pursuing these issues will help us to consolidate our appreciation of the balance of diversity and similarity within the broad field of *Leadership and Local Power in European Rural Development*.

Note

1   We draw here on the ideas presented in an unpublished discussion document produced for the COST group by Henri Goverde, Henk de Haan and Mireia Baylina, entitled 'Rurality in the Face of Gender and Power'.

References

Blomley, N. (1994), *Law, Space and the Geographies of Power*, Guilford Press, New York.

Ellis, A. (forthcoming), *Power and Participation in Grassroots Rural Development: The Case of LEADER II in Wales*, PhD thesis, University of Wales Swansea.

Goodwin, M. (1998), 'The Governance of Rural Areas: Some Emerging Research Issues and Agendas', *Journal of Rural Studies*, vol. 14, pp. 5–12.

Hoggart, K., Buller, H. and Black, R. (1995), *Rural Europe: Identity and Change*, Arnold, London.

Jessop, B. (1994), 'Post-Fordism and the State', in A. Amin (ed.), *Post-Fordism: A Reader*, Blackwell, Oxford, pp. 251–79.

Murdoch, J. (1995), 'Actor-Networks and the Evolution of Economic Forms: Combining Description and Explanation in Theories of Regulation, Flexible Specialization, and Networks', *Environment and Planning A*, vol. 27, pp. 731–57.

Murdoch, J. (2000), 'Networks – A New Paradigm of Rural Development?', *Journal of Rural Studies*, vol. 16, pp. 407–19.

Printed and bound by CPI Group (UK) Ltd, Croydon, CR0 4YY

21/10/2024

01777082-0006